Work Design
A Systematic Approach

Systems Innovation Series

Series Editor

Adedeji B. Badiru

Air Force Institute of Technology (AFIT) – Dayton, Ohio

PUBLISHED TITLES

Additive Manufacturing Handbook: Product Development for the Defense Industry, *Adedeji B. Badiru, Vhance V. Valencia, & David Liu*

Carbon Footprint Analysis: Concepts, Methods, Implementation, and Case Studies, *Matthew John Franchetti & Defne Apul*

Cellular Manufacturing: Mitigating Risk and Uncertainty, *John X. Wang*

Communication for Continuous Improvement Projects, *Tina Agustiady*

Computational Economic Analysis for Engineering and Industry, *Adedeji B. Badiru & Olufemi A. Omitaomu*

Conveyors: Applications, Selection, and Integration, *Patrick M. McGuire*

Culture and Trust in Technology-Driven Organizations, *Frances Alston*

Design for Profitability: Guidelines to Cost Effectively Management the Development Process of Complex Products, *Salah Ahmed Mohamed Elmoselhy*

Global Engineering: Design, Decision Making, and Communication, *Carlos Acosta, V. Jorge Leon, Charles Conrad, & Cesar O. Malave*

Global Manufacturing Technology Transfer: Africa–USA Strategies, Adaptations, and Management, *Adedeji B. Badiru*

Guide to Environment Safety and Health Management: Developing, Implementing, and Maintaining a Continuous Improvement Program, *Frances Alston & Emily J. Millikin*

Handbook of Construction Management: Scope, Schedule, and Cost Control, *Abdul Razzak Rumane*

Handbook of Emergency Response: A Human Factors and Systems Engineering Approach, *Adedeji B. Badiru & LeeAnn Racz*

Handbook of Industrial Engineering Equations, Formulas, and Calculations, *Adedeji B. Badiru & Olufemi A. Omitaomu*

Handbook of Industrial and Systems Engineering, Second Edition, *Adedeji B. Badiru*

Handbook of Military Industrial Engineering, *Adedeji B. Badiru & Marlin U. Thomas*

Industrial Control Systems: Mathematical and Statistical Models and Techniques, *Adedeji B. Badiru, Oye Ibidapo-Obe, & Babatunde J. Ayeni*

Industrial Project Management: Concepts, Tools, and Techniques, *Adedeji B. Badiru, Abidemi Badiru, & Adetokunboh Badiru*

Inventory Management: Non-Classical Views, *Mohamad Y. Jaber*

Kansei Engineering—2-volume set

- Innovations of Kansei Engineering, *Mitsuo Nagamachi & Anitawati Mohd Lokman*
- Kansei/Affective Engineering, *Mitsuo Nagamachi*

Kansei Innovation: Practical Design Applications for Product and Service Development, *Mitsuo Nagamachi & Anitawati Mohd Lokman*

Knowledge Discovery from Sensor Data, *Auroop R. Ganguly, João Gama, Olufemi A. Omitaomu, Mohamed Medhat Gaber, & Ranga Raju Vatsavai*

Learning Curves: Theory, Models, and Applications, *Mohamad Y. Jaber*

Managing Projects as Investments: Earned Value to Business Value, *Stephen A. Devaux*

PUBLISHED TITLES

Modern Construction: Lean Project Delivery and Integrated Practices, *Lincoln Harding Forbes & Syed M. Ahmed*

Moving from Project Management to Project Leadership: A Practical Guide to Leading Groups, *R. Camper Bull*

Profit Improvement through Supplier Enhancement, *Ralph R. Pawlak*

Project Feasibility: Tools for Uncovering Points of Vulnerability, *Olivier Mesly*

Project Management: Systems, Principles, and Applications, *Adedeji B. Badiru*

Project Management for the Oil and Gas Industry: A World System Approach, *Adedeji B. Badiru & Samuel O. Osisanya*

Project Management for Research: A Guide for Graduate Students, *Adedeji B. Badiru, Christina Rusnock, & Vhance V. Valencia*

Project Management Simplified: A Step-by-Step Process, *Barbara Karten*

Quality Management in Construction Projects, *Abdul Razzak Rumane*

Quality Tools for Managing Construction Projects, *Abdul Razzak Rumane*

A Six Sigma Approach to Sustainability: Continual Improvement for Social Responsibility, *Holly A. Duckworth & Andrea Hoffmeier*

Social Responsibility: Failure Mode Effects and Analysis, *Holly Alison Duckworth & Rosemond Ann Moore*

Statistical Techniques for Project Control, *Adedeji B. Badiru & Tina Agustiady*

STEP Project Management: Guide for Science, Technology, and Engineering Projects, *Adedeji B. Badiru*

Sustainability: Utilizing Lean Six Sigma Techniques, *Tina Agustiady & Adedeji B. Badiru*

Systems Thinking: Coping with 21st Century Problems, *John Turner Boardman & Brian J. Sauser*

Techonomics: The Theory of Industrial Evolution, *H. Lee Martin*

Total Productive Maintenance: Strategies and Implementation Guide, *Tina Agustiady & Elizabeth A. Cudney*

Total Project Control: A Practitioner's Guide to Managing Projects as Investments, Second Edition, *Stephen A. Devaux*

Triple C Model of Project Management: Communication, Cooperation, Coordination, *Adedeji B. Badiru*

Work Design: A Systematic Approach, *Adedeji B. Badiru & Sharon C. Bommer*

Work Design
A Systematic Approach

By

Adedeji B. Badiru
Sharon C. Bommer

CRC Press is an imprint of the
Taylor & Francis Group, an **informa** business

CRC Press
Taylor & Francis Group
6000 Broken Sound Parkway NW, Suite 300
Boca Raton, FL 33487-2742

Printed on acid-free paper

International Standard Book Number-13: 978-1-4987-5573-3 (Paperback)
International Standard Book Number-13: 978-1-1387-3176-9 (Hardback)

Library of Congress Cataloging-in-Publication Data

Names: Badiru, Adedeji Bodunde, 1952- editor. | Bommer, Sharon C., editor.
Title: Work design : a systematic approach / [edited by] Adedeji B. Badiru, Sharon C. Bommer.
Description: Boca Raton, FL : CRC Press, 2017. | Series: Industrial innovation series | Includes index.
Identifiers: LCCN 2016051396| ISBN 9781498755733 (pbk. : alk. paper) | ISBN 9781498755740 (ebook)
Subjects: LCSH: System analysis. | Systems engineering. | Industrial engineering. | Workflow.
Classification: LCC T57.6 .W667 2017 | DDC 658.3/01--dc23
LC record available at https://lccn.loc.gov/2016051396

Visit the Taylor & Francis Web site at
http://www.taylorandfrancis.com

and the CRC Press Web site at
http://www.crcpress.com

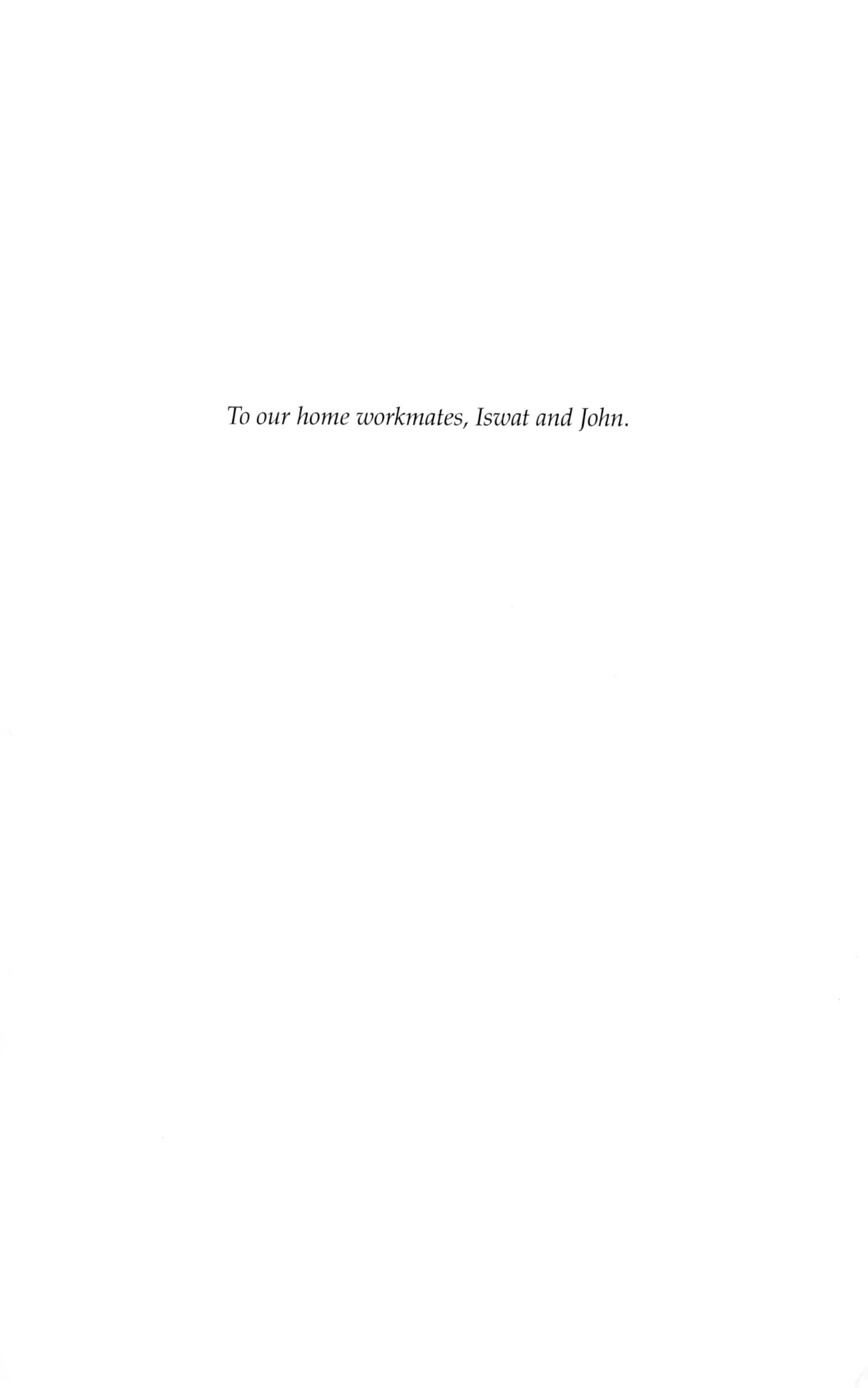

To our home workmates, Iswat and John.

Contents

Section IV: Work justification

Section V: Work integration

Preface

A systems view of work requires a systematic approach to managing work. This is exactly what this book provides. Work is all around us. Work permeates everything we do. Everyday activities such as writing, producing, gardening, building, cleaning, cooking, and so on are all around us. Not all work is justified, not all work is properly designed, not all work is evaluated accurately, and not all work is integrated. A systems model, as presented in *Work Design: A Systematic Approach*, will make work more achievable through better management. Work is defined as the process of performing a defined task or activity, such as research, development, operations, maintenance, repair, assembly, production, administration, sales, software development, inspection, data collection, data analysis, teaching, and so on. In essence, work defines everything we do. In spite of its being ubiquitous, very little explicit guidance is available in the literature on how to design, evaluate, justify, and integrate work. A unique aspect of this book is the inclusion of the cognitive aspects of work performance. Through a comprehensive systems approach, this book facilitates a better understanding of work for the purpose of making it more effective and rewarding. Topics covered include the definition of work, the workers' hierarchy of needs vis-à-vis the organization's hierarchy of needs, work performance measurement, conceptual frameworks for work management, analytic tools for work performance assessment, work design, work evaluation, work justification, work integration, and work control.

Acknowledgments

We thank our colleagues at the Air Force Institute of Technology (AFIT) and Wright State University for their continual input and suggestions, which have enhanced the quality of this manuscript. We are particularly grateful to the library staff at AFIT, with whom we did crowdsourcing for the selection of the book-cover design from the three options presented by the publisher's designers. We believe crowdsourcing the input of librarians is a first in the publishing industry, and we are delighted that this book on "work design" may set a work-involvement trend to be emulated by other authors. Special thanks go to our spouses, Iswat Badiru and John Bommer, who endured several months of intermittent disconnection while we sequestered ourselves writing, rewriting, and revising segments of the manuscript.

Authors

Adedeji B. Badiru is a professor of systems engineering at the Air Force Institute of Technology. He is a registered professional engineer, a fellow of the Institute of Industrial and Systems Engineers, and a fellow of the Nigerian Academy of Engineering. He is also a certified project management professional. He has a PhD in industrial engineering from the University of Central Florida. Dr. Badiru is the author of several books and technical journal articles. His areas of interest include project management, mathematical modeling, computer simulation, learning curve analysis, quality engineering, and productivity improvement.

Sharon C. Bommer is the Deputy Director of the Center for Operational Sciences (COA) at the Air Force Institute of Technology, and also an adjunct professor of industrial engineering in the Department of Biomedical, Industrial, and Human Factors Engineering at Wright State University. She has a PhD in industrial and human systems with specialization in human performance and cognition. Her areas of expertise include lean manufacturing, program management, and human systems integration. Dr. Bommer has extensive practical experience in industry, with more than fifteen years of automotive manufacturing engineering and operation experience.

section one

Systems overview

chapter one

Systems view of work

> In teamwork we work and thrive so that our work works well.
>
> **Adedeji Badiru,**
> *2016*

Poor work design can happen at any level of an organization, whether at the blue collar, white collar, or gold collar level (as author Adedeji Badiru refers to top executives). This point was driven home by a young technical executive who sent a message home, saying, "Sorry I wasn't able to complete the family errand today. Work has been crazy. I am behind on emails and everything else. I have eleven straight weeks of traveling coming up. I hope to complete the errand as soon as I can." This is a case of work interfering with family need. The family should be a part of the overall work system design for an effective and sustainable execution of work. A systems view facilitates the much-desired work-life balance.

The premise of *Work Design: A Systematic Approach* is to combine the respective systems engineering backgrounds and cognitive expertise of the authors into a comprehensive treatment of how work is designed, evaluated, justified, and integrated, based on the systems engineering model of *design*, *evaluation*, *justification*, and *integration* (DEJI™) (Badiru, 2014). The book is presented in five sections: (1) systems view of work, (2) work design, (3) work evaluation, (4) work justification, and (5) work integration.

Work design: The planning, organizing, and coordination of work elements all fall under the category of design in the DEJI™ model. It is essential that a structured approach be applied to work design right from the outset. Retrofitting a work element only after problems develop not only impedes the overall progress of work in an organization, but it also leads to an inefficient use of limited human and material resources. This stage of the model is expected to guide work designers onto the path of strategic thinking about work elements down the line rather than just the tactical manipulation of work for the present needs. In this regard, Badiru (2016) says, "Right next to innovation, structured methods for producing effective work results are a survival imperative for every large organization."

Work evaluation: Following the design of a work element, the DEJI model calls for a formal evaluation of the intended purpose of the work vis-à-vis other work elements going on in the organization. Such an evaluation may lead to a need to go back and redesign the work element. Evaluation can be done as a combination of both qualitative and quantitative assessment of the work element, depending on the specific nature of the work, the main business of the organization, and the managerial capabilities of the organization.

Work justification: According to the concept of the DEJI model, not only should a work element be designed and evaluated, it should also be formally and rigorously justified. If this is not done, errant work elements will creep into the organizational pursuits. What is worth doing is worth doing well; otherwise, it should not be done at all. The principles of *lean* operations (Agustiady and Badiru, 2012) suggest weeding out functions that do not add value to the organizational goal. In this regard, each and every work element needs to be justified. But it should be realized that not all work elements are expected to generate physical products in the work environment. A work element may be justified on the basis of adding value to the well-being of the worker with respect to his or her mental, emotional, spiritual, and physical characteristics. The point of this stage of the DEJI model is to determine whether the work element is needed at all. If not, the do-nothing alternative is always an option.

Work integration: This last stage of the DEJI model is of the utmost importance, but it is often neglected. The model affirms that the most sustainable work elements are those that fit within the normal flow of operations, existing practices, or other expectations within an organization. Does the work fit in? Will a new work element under consideration be an extraneous pursuit or a detraction from the overall work plan? If a work element is not integrated with other normal pursuits, it cannot be sustained for the long haul. This is why many organizations suffer from repeated program starts and stops. For example, in as much as worker wellness programs are desirable pursuits in an organization, they cannot be sustained if they are not integrated into the culture and practices of the organization. Unintegrated flash-in-the-pan programs, activities, and work elements often fall by the wayside over time. For this stage of the DEJI model, work elements must be tied to the end goal of the person and the organization. The DEJI model suggests that, before work starts, we should always explore the easiest, fastest, most efficient, and most effective way to execute the work. Badiru (2016) remarks that "aimless work is so insidious because it tends to covertly masquerade as fruitful labor," which is not connected to real organizational goals. This remark also suggests distinguishing between doing real work and creating the illusion of working—that is, doing productive work versus doing busy work without a tangible output. This categorization of work is the basis for the premise of applying the multidimensional hierarchy of needs (of the worker and the organization), as discussed in a subsequent section of this chapter.

Work is the means to accomplish a goal. For the purpose of this book, work is literally an activity to which strength, mental acuity, and resources are applied to get something done. This may involve a sustained physical and/or mental effort to overcome impediments in the pursuit of an outcome, an objective, a result, or a product. In an operational context, work can be viewed as the process of performing a defined task or activity, such as research, development, operations, maintenance, repair, assembly, production, administration, sales, software development, inspection, data collection, data analysis, teaching, and so on. The opening quote in this chapter signifies the meaning of work as a means to an end goal, where workers thrive and the work effort succeeds. If you understand your work from a systems viewpoint, you will enjoy the work and you will want to do more of it. In this book, we view work as a *work system* rather than work in isolation or disconnected from other human endeavors.

A systems view of work is essential because of the several factors and diverse people who may be involved in the performance of the work. There are systems and subsystems involved in the execution of work whether small or large, simple or complex, localized or multilocational. For an activity to be consistently workable, all the attendant factors and issues must be taken into account. If some crucial factors are neglected, the *workability* of the activity may be in jeopardy. Geisler (2012) highlights the important elements for working happy and doing happy work. These elements include

- Work leadership
- Work rules and regulations
- Work flow
- Work space
- Work teams
- Work duties and responsibilities
- Work compensation
- Work fulfillment

Only a systems viewpoint can adequately capture the multitude of diverse elements involved in working happy. People work better when they are happy. According to author Adedeji Badiru, these are "happy-go-working" people. Systems-based analysis can facilitate the optimal design of work within the prevailing scenario in the work environment. In many cases, it is not necessary to work harder; it is just essential to design the work better.

For the purpose of a systems view of work, we define a system as a collection of interrelated elements (subsystems) working together synergistically to generate a collective and composite outcome (value) that is higher than the sum of the individual outcomes of the subsystems. Even simple tasks run the risk of failure if some minute subsystem is not accounted for. The following two specific systems engineering models are used for the purpose of this book.

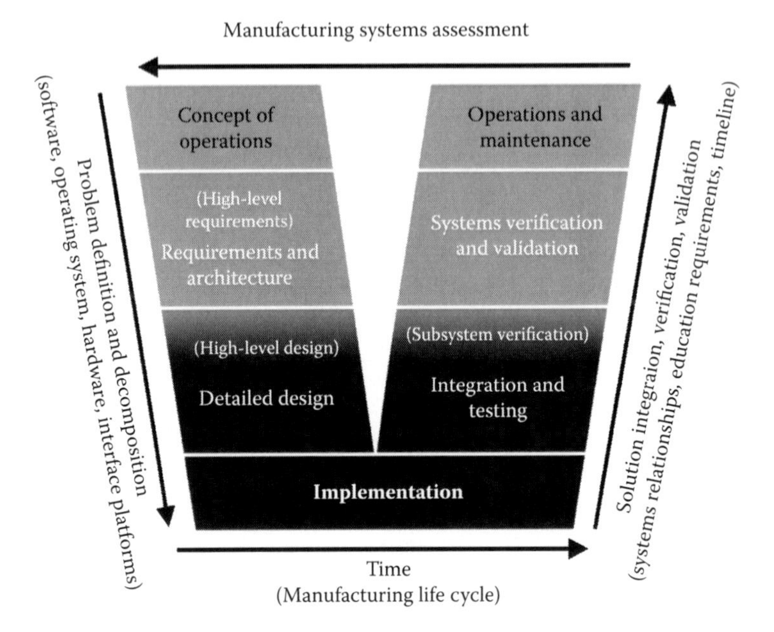

Figure 1.1 An example V-model of systems engineering applied to a manufacturing system.

1. The V-model of systems engineering
2. The DEJI model of systems quality integration (Badiru, 2014)

Figure 1.1 presents an illustration of the V-model applied to a manufacturing enterprise consisting of a series of work elements. Although the model is most often used in software development processes, it is also applicable to hardware development as well as general work in systems engineering. In the model, instead of moving down in a linear fashion, the process steps are bent upward after the coding phase to form the typical V shape. The V-model demonstrates the relationships between each phase of the development life cycle and its associated phase of testing. The horizontal and vertical axes represent time or project completeness (left to right) and the level of abstraction, respectively.

There are several different ways that a work system can be developed and delivered using the V-model. The best development strategy depends on how much the work analyst knows about the system for which the work is being designed. Three basic design strategies can be used.

Once-through approach: In this case, we plan, specify, and implement the complete work system in one pass through the V shape. This approach, also sometimes called the *waterfall approach*, works well if the vision is clear, the requirements are well understood and stable, and there is sufficient funding.

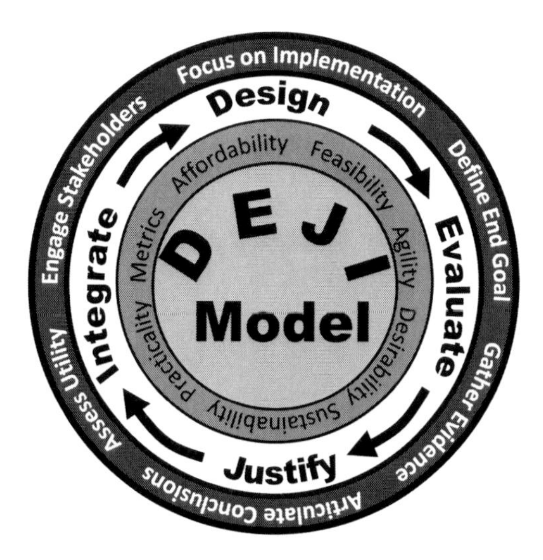

Figure 1.2 Framework for the application of the DEJI™ model to systems integration.

The problem is that there isn't a lot of flexibility or opportunity for recovery if the vision, work environment, or requirements change substantially.

Incremental approach: Here, we plan and specify the work system and then implement it in a series of well-defined increments or phases, where each increment delivers a portion of the desired end goal. This is like moving through value-adding increments of the work. In this case, we are making one pass through the first part of the V shape and then iterating through the latter part for each phased-in increment. This is a common strategy for field equipment deployment, where system requirements and design can be incrementally implemented and deployed across a given area in several phases and several projects.

Evolutionary approach: In this approach, we plan, specify, and implement an initial system capability, learn from the experience with the initial system, and then define the next iteration to address issues and extend capabilities (or add value). Thus, we refine the *concept of operations*, add and change system requirements, and revise the design as necessary. We will continue with successive iterative refinements until the work system is complete. This strategy can be shown as a series of Vs that are placed end to end, since system operation on the right side of the V influences the next iteration. This strategy provides the most flexibility but also requires project management expertise and vigilance to make sure that the development stays on track. It also requires patience from the stakeholders as the design moves along in incremental stages.

Figure 1.2 presents an illustration of the application of the DEJI model for the systems integration of work factors. The key benefit of the DEJI model is that it moves the effort systematically through the stages of design, evaluation, justification, and integration. The approach queries the work analyst about what needs to be addressed at each stage so that he or she does not drop the ball on critical requirements. The greatest aspect of the DEJI model is the final stage, which calls for the integration of the work with other efforts within the work environment. If there is a disconnect, then the work may end up being a misplaced effort. If a work effort is properly integrated, then it will be sustainable.

Systems definition

A system is defined as a collection of interrelated elements working together synergistically to achieve a set of objectives. Any work is essentially a collection of interrelated activities, tasks, people, tools, resources, processes, and other assets brought together in the pursuit of a common goal. The goal may be set in terms of generating a physical product, providing a service, or achieving a specific result. This makes it possible to view any work as a system that is amenable to all the classical and modern concepts of systems management.

Work is the foundation of everything we do. Having some knowledge is not enough; the knowledge must be applied to the pursuit of objectives. Work management facilitates the application of knowledge and the willingness to actually accomplish tasks. Where there is knowledge, willingness often follows. But it is the execution of the work that actually gets things done. From very basic tasks to very complex endeavors, work management must be applied. It is thus essential that systems thinking be a part of the core of every work pursuit in business, industry, education, government, and even at home. In this regard, a systems approach is of the utmost importance because work accomplishment is a "team sport" that has several underlying factors as elements of the overall work system.

We thrive together when we showcase our work together. Also, working together productively requires that the work be designed appropriately to permit teamwork from a systems perspective.

Technical systems control

Classical technical systems control focuses on the control of the dynamics of mechanical objects, such as pumps, electrical motors, turbines, rotating wheels, and so on. The mathematical basis for such control systems can be adapted (albeit in iconic formats) to management systems, including those for work management. This is because both technical and managerial systems are characterized by inputs, variables, processing, control, feedback, and output. This is represented graphically by input–process–output relationship block diagrams. Mathematically, it can be represented as

$$z = f(x) + \varepsilon$$

where:

z = output
$f(.)$ = functional relationship
ε = error component (noise, disturbance, etc.)

For multivariable cases, the mathematical expression is represented by vector–matrix functions, as follows:

$$Z = f(x) + E$$

where each term is a matrix:

Z = output vector
$f(x)$ = input vector
E = error vector

Regardless of the level or form of mathematics used, all systems exhibit the same input–process–output characteristics, either quantitatively or qualitatively. The premise of this book is that there should be a cohesive coupling of quantitative and qualitative approaches in managing a work system. In fact, it is this unique blend of approaches that makes the systems application for work management more robust than what one finds in mechanical control systems, where the focus is primarily on quantitative representations.

Improved organizational performance

Systems engineering efficiency and effectiveness are of interest across the spectrum of work management for the purpose of improving organizational performance. Managers, supervisors, and analysts should be interested in having systems engineering serve as the umbrella for improving work efforts throughout the organization. This will properly connect everyone with the prevailing organizational goals as well as create collaborative avenues among the personnel. Systems application applies across the spectrum of any organization and encompasses the following elements:

- Technological systems (e.g., engineering control systems and mechanical systems)
- Organizational systems (e.g., work process design and operating structures)
- Human systems (e.g., interpersonal relationships and human–machine interfaces)

A systems view of the world makes everything work better and work efforts more likely to succeed. A systems view provides a disciplined process for the design, development, and execution of work both in technical and nontechnical organizations. One of the major advantages of a systems approach is the win–win benefit for everyone. A systems view also allows the full involvement of all stakeholders and constituents of a work center. This is articulated well by the following saying:

Tell me and I forget;
Show me and I remember;
Involve me and I understand.
(Confucius, Chinese philosopher)

For example, from a systems perspective, the pursuit of organizational or enterprise transformation is best achieved through the involvement of everyone. Every work environment is complex because of the diversity of factors involved, including the following:

- Workers' overall health and general well-being
- Workers' physical attributes
- Workers' mental abilities
- Workers' emotional stability
- Workers' spiritual interests
- Workers' psychological profiles

There are differing human personalities. There are differing technical requirements. There are differing expectations. There are differing environmental factors. Each specific context and prevailing circumstances determine the specific flavor of what can and cannot be done in the work environment. The best approach for effective work management is to adapt to each work requirement and specification. This means taking a systems view of the work. The work systems approach presented in this book is needed for *working across* organizations, countries, cultures, and the unique nuances of each project. This is an essential requirement in today's globalized and intertwined personal and professional environments. A systems view requires embracing multiple disciplines in the execution of work in a way that each component complements others in the system. Formal work management represents an excellent platform for the implementation of a systems approach. A comprehensive work management program requires control techniques such as operations research, operations management, forecasting, quality control, and simulation to achieve goals. Traditional approaches to management use these techniques in a disjointed fashion, thus ignoring the potential interplay among the techniques. The need for integrated systems-based work management worldwide has been recognized for decades. As long ago as 1993, the World Bank reported that a lack of systems accountability led to several global project failures. The bank, which has loaned more than $300 billion to developing countries over the last half-century, acknowledged that there has been a dramatic rise in the number of failed projects around the world. In other words, the work efforts failed. The lack of an integrated systems approach to managing the projects was cited as one of the major causes of failure. More recent reports by other organizations point to the same flaws in managing global projects and to the need to apply better project management to major projects. A Government Accountability Office (GAO) report, publicly released in 2010 (GAO, 2010), highlights that defense needs better management of projects. This was in the wake of a government audit that revealed gross inefficiencies in managing large defense projects. In a national news release on April 1, 2008, it was reported that auditors at the Government Accountability Office (GAO) issued a scathing review of dozens of the Pentagon's biggest weapons systems, citing that ships, aircraft, and satellites are billions of dollars over budget and years behind schedule. According to the review, "95 major systems have exceeded their original budgets by a total of $295 billion; and are delivered almost two years late on average." Further, "none of the systems that the GAO looked at had met all of the standards for best management practices during their development stages." Among the programs noted for increased development costs were the Joint Strike Fighter and Future Combat Systems. The costs of those programs had risen 36% and 40%, respectively, while C-130 avionics modernization costs had risen 323%. And, while "Defense Department officials have tried to improve the procurement process," the GAO added that "significant policy changes have not yet translated into best practices on individual programs." In our view, a failed program is an indicator of failed work efforts. A summary of the report (GAO, 2010) reads as follows:

> Every dollar spent inefficiently in developing and procuring weapon systems is less money available for many other internal and external budget priorities, such as the global war on terror and growing entitlement programs. These inefficiencies also often result in the delivery of less capability than initially planned, either in the form of fewer quantities or delayed delivery to the warfighter.

In as much as the military represents the geo-politico-economic landscape of a nation, the preceding assessment is representative of what every organization faces, whether public or private. In systems-based project management, it is essential that related techniques be employed in an integrated fashion so as to maximize the total project output. One definition of systems project management is stated as follows:

> Systems project management is the process of using systems approach to manage, allocate, and time

> resources to achieve systems-wide goals in an efficient and expeditious manner. (Badiru, 2012)

This definition calls for the systematic integration of technology, human resources, and work process design to achieve goals and objectives. There should be a balance in the synergistic integration of humans and technology. There should not be an overreliance on technology, nor should there be an overdependence on human processes. Similarly, there should not be too much emphasis on analytical models to the detriment of commonsense human-based decisions.

What is systems engineering?

Systems engineering is growing in appeal as an avenue to achieve organizational goals and improve operational effectiveness and efficiency. Researchers and practitioners in business, industry, and government all embrace systems engineering implementations. So, what is systems engineering? Several definitions exist, but the one offered below is quite comprehensive.

Systems engineering is the application of engineering to solutions of a multifaceted problem through a systematic collection and integration of parts of the problem with respect to the lifecycle of the problem. It is the branch of engineering concerned with the development, implementation, and use of large or complex systems. It focuses on specific goals of a system considering the specifications, prevailing constraints, expected services, possible behaviors, and structure of the system. It also involves a consideration of the activities required to assure that the system's performance matches the stated goals. Systems engineering addresses the integration of tools, people, and processes required to achieve a cost-effective and timely operation of the system.

The International Council on Systems Engineering (INCOSE) defines systems engineering as follows:

> Systems Engineering is an interdisciplinary approach and means to enable the realization of successful systems. It focuses on defining customer needs and required functionality early in the development cycle, documenting requirements, then proceeding with design synthesis and system validation while considering the complete problem. (INCOSE, 2017)

Systems engineering integrates all the disciplines and specialty groups into a team effort, forming a structured development process that proceeds from concept to production to operation. Systems engineering considers both the business and the technical needs of all involved with the organizational goals.

Work systems logistics

Logistics can be defined as the planning and implementation of a complex task, the planning and control of the flow of goods and materials through an organization or manufacturing process, or the planning and organization of the movement of personnel, equipment, and supplies. Complex projects represent a hierarchical system of operations. Thus, we can view a project system as a collection of interrelated projects all serving a common end goal. Consequently, we present the following universal definition:

> Work systems logistics is the planning, implementation, movement, scheduling, and control of people, equipment, goods, materials, and supplies across the interfacing boundaries of several related projects.

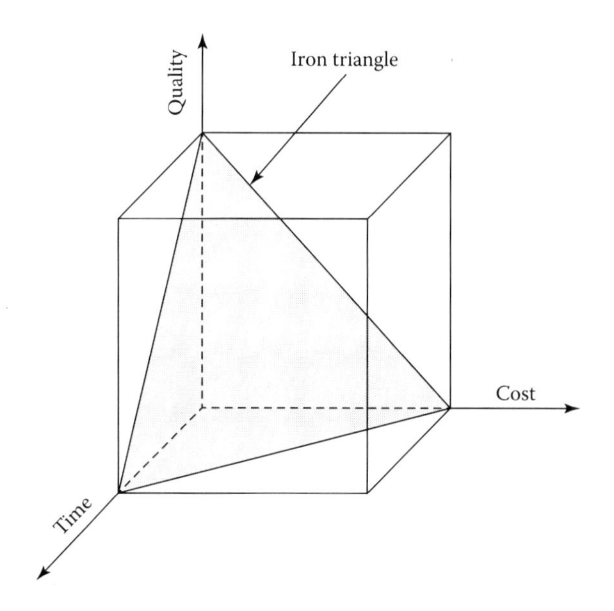

Figure 1.3 Systems constraints of cost, time, and quality within an iron triangle.

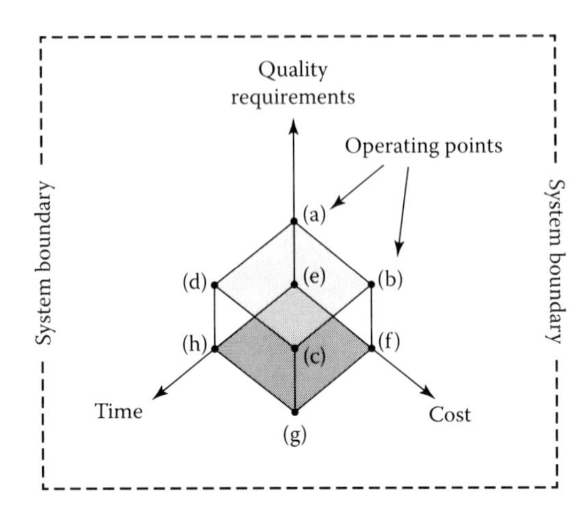

Figure 1.4 Systems constraints of cost, time, and quality within boxed system boundaries.

Conventional organizational management must be modified and expanded to address the unique logistics of work systems.

Systems constraints

Systems management is the pursuit of organizational goals within the constraints of time, cost, and quality expectations. The iron triangle model depicted in Figure 1.3 shows that project accomplishments are constrained by the boundaries of quality, time, and cost. In this case, quality represents the composite collection of project requirements. In a situation where precise optimization is not possible, there will have to be trade-offs between these three factors of success. The concept of the iron triangle is that a rigid triangle of constraints encases the project. Everything must be accomplished within the boundaries of time, cost, and quality. If better quality is expected, a compromise along the axes of time and cost must be executed, thereby altering the shape of the triangle. Trade-off relationships are not linear and must be visualized in a multidimensional context. This is better articulated by the alternative view of the systems constraints shown in Figure 1.4. Scope requirements determine the project boundary, and trade-offs must be made within that boundary. If we label the eight corners of the box (a) through (h), we can iteratively assess the best operating point for the project. For example, we can address the following two operational questions:

1. From the point of view of the project sponsor, which corner is the most desired operating point in terms of combination of requirements, time, and cost?

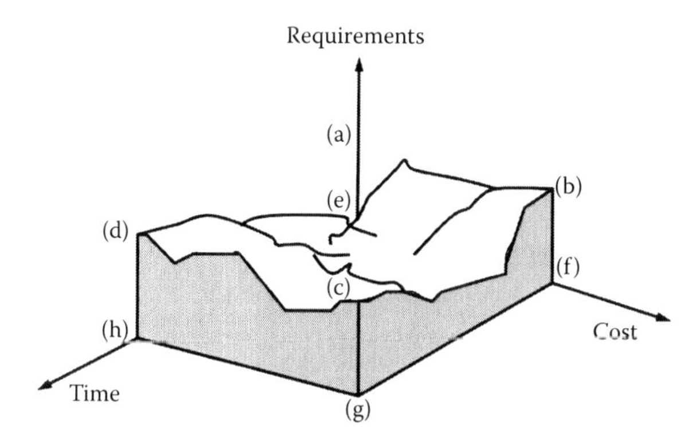

Figure 1.5 Compromise surface for cost, time, and requirements trade-off.

2. From the point of view of the project executor, which corner is the most desired operating point in terms of combination of requirements, time, and cost?

Note that all the corners represent extreme operating points. We notice that point (e) is the do-nothing state, where there are no requirements, no time allocation, and no cost incurrences. This cannot be the desired operating state of any organization that seeks to remain productive. Point (a) represents an extreme case of meeting all requirements with no investment of time or cost allocation. This is an unrealistic extreme in any practical environment. It represents a case of getting something for nothing. Yet, it is the most desired operating point for the project sponsor. By comparison, point (c) provides the maximum possible requirements, cost, and time. In other words, the highest levels of requirements can be met if the maximum possible time is allowed and the highest possible budget is allocated. This is an unrealistic expectation in any resource-conscious organization. You cannot get everything you ask for to execute a project. Yet, it is the most desired operating point for the project executor. Considering the two extreme points of (a) and (c), it is obvious that the project must

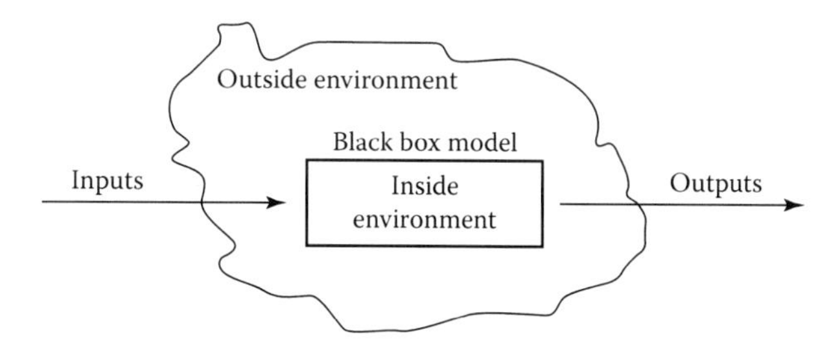

Figure 1.6 Outside vs. inside environments of a work system.

be executed within some compromise region within the scope boundary. Figure 1.5 shows a possible view of a compromise surface, with peaks and valleys representing give-and-take trade-off points within the constrained box. The challenge is to come up with an analytical modeling technique to guide decision making over the compromise region. If we could collect sets of data over several repetitions of identical projects, then we could model a decision surface that can guide the executions of similar future projects. Such typical repetitions of an identical project are most readily apparent in construction—for example, residential home development projects.

Systems framework for work effectiveness

The philosophy of systems influence suggests that you control the internal environment while only influencing the external environment. In Figure 1.6, the inside (controllable) environment is represented as a black box in the typical input–process–output relationship. The outside (uncontrollable) environment is bounded by the cloud representation. In the comprehensive systems structure, inputs come from the global environment, are moderated by the immediate outside environment, and are delivered to the inside environment. In an unstructured inside environment, work functions occur as blobs, as illustrated in Figure 1.7. A "blobby" environment is characterized by intractable activities where everyone is busy but without a cohesive structure of input–output relationships. In such a case, the following disadvantages may be present.

- Lack of traceability
- Lack of process control

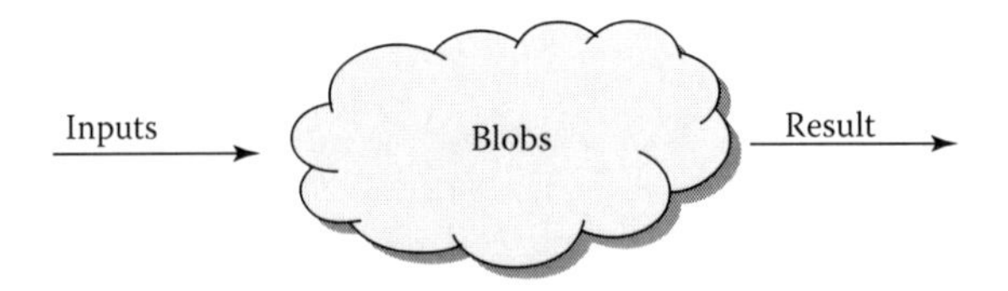

Figure 1.7 Blobs of work in unstructured programs.

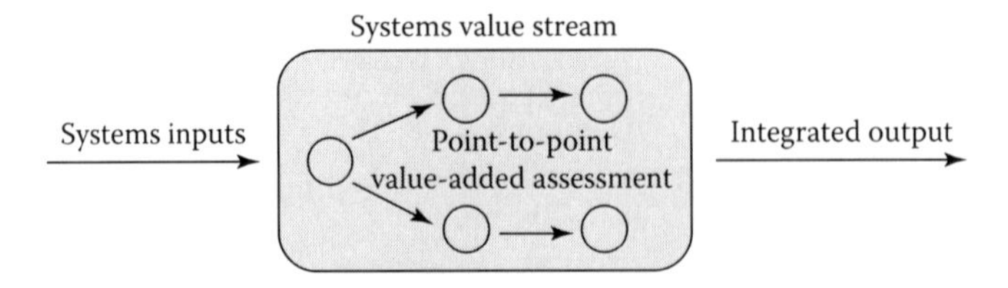

Figure 1.8 Systems value stream structure for work enhancement.

- Higher operating costs
- Inefficient personnel interfaces
- Unrealized technology potentials

Organizations often inadvertently fall into the blobby structure because it is simple, low cost, and less time consuming—until a problem develops. A desired alternative is to model the project system using a systems value stream structure, as shown in Figure 1.8. This uses a proactive and problem-preempting approach to executing projects. This alternative has the following advantages:

- Problem diagnosis is easier.
- Accountability is higher.
- Operating waste is minimized.
- Conflict resolution is faster.
- Value points are traceable.

Work systems value modeling

A technique that can be used to assess the overall value-added components of a process improvement program is the *systems value model* (SVM), which is an adaptation of the *manufacturing system value* (MSV) model presented by Troxler and Blank (1989). The model provides an analytical decision aid for comparing process alternatives. Value is represented as a p-dimensional vector, as follows:

$$V = f\left(A_1, A_2, \ldots, A_p\right)$$

where $A = (A_1, \ldots, A_n)$ is a vector of quantitative measures of tangible and intangible attributes.

Work Value is a function of Work Attributes.

Examples of process attributes are quality, throughput, capability, productivity, cost, and schedule. Attributes are considered to be a combined function of factors, x_1, expressed as

$$A_k\left(x_1, x_2, \ldots, x_{m_k}\right) = \sum_{i=1}^{m_k} f_i\left(x_i\right)$$

where $\{x_i\}$ = the set of m factors associated with attribute $A_k (k = 1,2,\ldots,p)$ and f_i = the contribution function of factor x_i to attribute A_k.

Work Attribute is a function of Work Factors.

Examples of factors include reliability, flexibility, user acceptance, capacity utilization, safety, and design functionality. Factors are themselves considered to be composed of indicators, v_i, expressed as follows:

$$x_i\left(v_1, v_2, \ldots, v_n\right) = \sum_{j=1}^{n} z_i\left(v_i\right)$$

where $\{v_j\}$ = the set of n indicators associated with factor $x_i (i = 1,2,\ldots,m)$ and z_j = the scaling function for each indicator variable v_j.

Work Factor is a function of Work Indicators.

Examples of indicators are project responsiveness, lead time, learning curve, and work rejects. By combining the preceding definitions, a composite measure of the value of a process can be modeled as follows:

$$V = f\left(A_1, A_2, \ldots, A_p\right)$$

$$= f\left\{\left[\sum_{i=1}^{m_1} f_i\left(\sum_{j=1}^{n} z_j\left(v_j\right)\right)\right]_1, \left[\sum_{i=1}^{m_2} f_i\left(\sum_{j=1}^{n} z_j\left(v_j\right)\right)\right]_2, \ldots, \left[\sum_{i=1}^{m_k} f_i\left(\sum_{j=1}^{n} z_j\left(v_j\right)\right)\right]_p\right\}$$

where m and n may assume different values for each attribute.

Work Value is a function of Work Attributes, Work Factors, and Work Indicators.

A subjective measure to indicate the utility of the decision maker may be included in the model by using an attribute weighting factor, w_i, to obtain a weighted PV.

$$PV_w = f\left(w_1 A_1, w_2 A_2, \ldots, w_p A_p\right)$$

where:

$$\sum_{k=1}^{p} w_k = 1, \qquad \left(0 \le w_k \le 1\right)$$

With this modeling approach, a set of process options can be compared on the basis of a set of attributes and factors.

Table 1.1 Comparison of IT work value options

IT equipment options	Suitability ($k = 1$)	Capability ($k = 2$)	Performance ($k = 3$)	Productivity ($k = 4$)
Option A	0.12	0.38	0.18	0.02
Option B	0.30	0.40	0.28	−1.00
Option C	0.53	0.33	0.52	−1.10

Example of value vector modeling

To illustrate the model, suppose three information technology (IT) options are to be evaluated based on four attribute elements: *capability*, *suitability*, *performance*, and *productivity* (see Table 1.1). For this example, based on the relationships described above, the value vector is defined as follows:

$$V = f(\text{capability, suitability, performance, productivity})$$

Capability refers to the ability of IT equipment to satisfy multiple requirements. For example, a certain piece of IT equipment may only provide a computational service. A different piece of equipment may be capable of generating reports in addition to computational analysis, thus increasing the service variety that can be obtained. In Table 1.1, the levels of increase in service variety from the three competing equipment types are 38%, 40%, and 33%, respectively. *Suitability* refers to the appropriateness of the IT equipment for current operations. For example, the respective percentages of operating scope for which the three options are suitable are 12%, 30%, and 53%. *Performance* in this context refers to the ability of the IT equipment to satisfy schedule and cost requirements. In the example, the three options can, respectively, satisfy requirements on 18%, 28%, and 52% of the typical set of jobs. *Productivity* can be measured by an assessment of the performance of the proposed IT equipment to meet workload requirements in relation to the existing equipment. In the example in Table 1.1, the three options, respectively, show normalized increases of 0.02, −1.0, and −1.1 on a uniform scale of productivity measurement. A plot of the histograms of the respective "values" of the three IT options were evaluated to find option C as the best value alternative in terms of suitability and performance. Option B shows the best capability measure, but its productivity is too low to justify the required investment. Option A offers the best productivity, but its suitability measure is low. The analytical process can incorporate a lower control limit into the quantitative assessment such that any option providing value below that point will not be acceptable. Similarly, a minimum value target can be incorporated into the graphical plot such that each option is expected to exceed the target point on the value scale.

The relative weights used in many justification methodologies are based on the subjective propositions of decision makers. Some of those subjective weights can be enhanced by the incorporation of utility models. For example, the weights shown in Table 1.1 could be obtained from utility functions. There is a risk of spending too much time maximizing inputs at *point-of-sale* levels and too little time defining and refining outputs at the *wholesale* systems level. Without a systems view, we cannot be sure that we are pursuing the right outputs.

Application of work project management

Work should be viewed as a project. The field of project management continues to grow as an effective means of managing functions in any organization. Project management should be an enterprise-wide systems-based endeavor. This section presents a brief introduction to work project management. Chapter 10 presents full details of the tools and techniques of project mangement. Enterprise-wide project management is the application of project management techniques and practices across the full scope of the enterprise. This concept is also referred to as *management by project* (MBP). MBP is a contemporary concept that employs project management techniques in various functions within an organization. MBP recommends pursuing endeavors as project-oriented activities. It is an effective way to conduct any business activity. It represents a disciplined approach that defines any work assignment as a project. Under MBP, every undertaking is viewed as a project that must be managed just like a traditional project. The characteristics required of each project so defined are

1. An identified scope and a goal
2. A desired completion time
3. The availability of resources
4. A defined performance measure
5. A measurement scale for reviewing work

An MBP approach to operations helps to identify unique entities within functional requirements. This identification helps determine where functions overlap and how they are interrelated, thus paving the way for better planning, scheduling, and control. Enterprise-wide project management facilitates a unified view of organizational goals and provides a way for project teams to use information generated by other departments to carry out their functions.

The use of project management continues to grow rapidly. The need to develop effective management tools increases with the increasing complexity of new technologies and processes. The life cycle of a new product to be introduced into a competitive market is a good example of a complex process that must be managed with integrative project management approaches. The product will encounter management functions as it goes from one stage to the next. Project management will be needed throughout the design and production stages of the product. Project management will be needed in developing marketing, transportation, and delivery strategies for the product. When the product finally gets to the customer, project management will be needed to integrate its use with those of other products within the customer's organization.

The need for a project management approach is established by the fact that a project will always tend to increase in size even if its scope narrows. The following literary laws are applicable to any project environment.

Parkinson's law: Work expands to fill the available time or space.
Peter's principle: People rise to the level of their incompetence.
Murphy's law: Whatever can go wrong will.
Badiru's rule: The grass is always greener where you most need it to be dead.

An integrated systems project management approach can help diminish the adverse impacts of these laws through good project planning, organizing, scheduling, and control.

Integrated systems implementation

Project management tools can be classified into three major categories:

1. *Qualitative tools*: These are the managerial tools that aid in the interpersonal and organizational processes required for project management.
2. *Quantitative tools*: These are analytical techniques that aid in the computational aspects of project management.

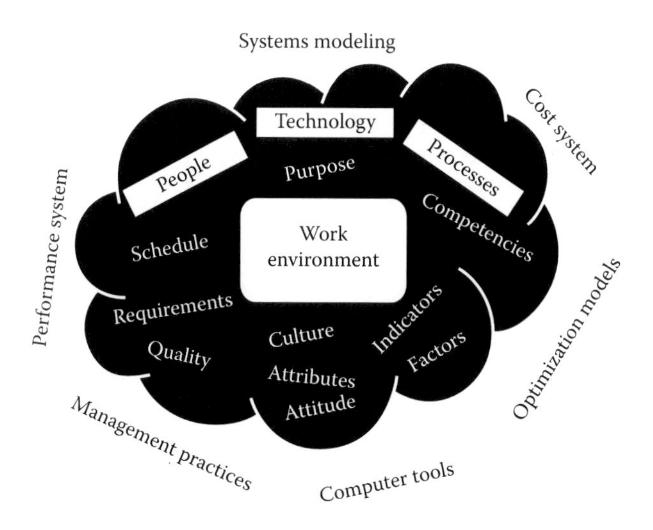

Figure 1.9 Project systems modeling in the work environment.

3. *Computer tools*: These are software and hardware tools that simplify the process of planning, organizing, scheduling, and controlling a project. Software tools can help in both the qualitative and quantitative analyses needed for project management.

Although individual books dealing with work design, cognitive modeling, management principles, optimization models, and computer tools are available (Konz and Johnson, 2004; IEBOK, 2017), there are few guidelines for the integration of the topics explicitly for work project management purposes. In this book, we integrate most of these topics for a comprehensive guide to work project management, from a systems perspective. Figure 1.9 illustrates this emphasis. The approach considers not only the management of the work itself but also the management of all the worker-related functions that support the work.

It is one thing to have a quantitative model, but it is a different thing to be able to apply the model to real-world problems in a practical form. The systems approach presented in this book illustrates how to make the transition from model to practice.

A systems approach helps to increase the intersection of the three categories of project management tools and, hence, improve overall management effectiveness. Crisis should not be the instigator for the use of project management techniques. Project management approaches should be used upfront to prevent avoidable problems rather than to fight them when they develop. What is worth doing is worth doing well, right from the beginning.

Worker-based factors for systems success

The premise of this book is that the critical factors for systems success revolve around people and the personal commitment and dedication of each person. No matter how good a technology is and no matter how enhanced a process might be, it is ultimately the people involved who determine its success. This makes it imperative to take care of people issues first in the overall systems approach to work project management. Many organizations recognize this, but few are able to actualize the ideals of managing work and people productively. The execution of operational strategies requires forthrightness, openness, and commitment to get things done. Lip service and arm waving are not sufficient. Tangible programs that cater to the needs of people must be implemented. It is essential to provide incentives, encouragement, and empowerment for people to be self-actuating in determining how best to accomplish their job functions. A summary of critical factors for systems success encompasses total management of hardware, software, people, and processes with a particular focus on the following system attributes:

- Operational effectiveness
- Operational efficiency
- System suitability
- System resilience
- System affordability
- System supportability
- System life cycle cost
- System performance
- System schedule
- System cost

Systems engineering tools, techniques, and processes are essential for project life cycle management to make goals possible within the context of SMART principles, which are represented as follows:

1. *Specific*: Pursue specific and explicit outputs.
2. *Measurable*: Design outputs that can be tracked, measured, and assessed.
3. *Achievable*: Make outputs that are achievable and aligned with organizational goals.
4. *Realistic*: Pursue only the goals that are realistic and results oriented.
5. *Timed*: Make outputs timed to facilitate accountability.

Systems engineering provides the technical foundation for executing a project successfully. A systems approach is particularly essential in the early stages of the project in order to avoid having to reengineer the project at the end of its life cycle. Early systems engineering makes it possible to proactively assess the feasibility of meeting user needs, the adaptability of new technology, and the integration of solutions into regular operations.

Systems hierarchy

The traditional concepts of systems analysis are applicable to the project process. The definitions of a project system and its components follow.

System: A project system consists of interrelated elements organized for the purpose of achieving a common goal. The elements are organized to work synergistically to generate a unified output that is greater than the sum of the individual outputs of the components.

Program: A program is a very large and prolonged undertaking. Such endeavors often span several years. Programs are usually associated with particular systems. For example, we may have a space exploration program within a national defense system.

Project: A project is a time-phased effort of much smaller scope and duration than a program. Programs are sometimes viewed as consisting of a set of projects. Government projects are often called *programs* because of their broad and comprehensive nature. Industry tends to use the term *project* because of the short-term and focused nature of most industrial efforts.

Task: A task is a functional element of a project. A project is composed of a sequence of tasks that all contribute to the overall project goal.

Activity: An activity can be defined as a single element of a project. Activities are generally smaller in scope than tasks. In a detailed analysis of a project, an activity may be viewed as the smallest, most practically indivisible work element of the project. For example, we can regard a manufacturing plant as a system. A plant-wide endeavor to improve productivity can be viewed as a program. The installation of a flexible manufacturing system is a project within the productivity improvement program. The process of identifying and selecting equipment vendors is a task, and the actual process of placing an order with a preferred vendor is an activity.

The emergence of systems development has had an extensive effect on project management in recent years. A system can be defined as a collection of interrelated elements brought together to achieve a specified objective. In a management context, the purposes of a system are to develop and manage operational procedures and to facilitate an effective decision-making process. Some of the common characteristics of a system include

1. Interaction with the environment
2. Objectives
3. Self-regulation
4. Self-adjustment

Representative components of a project system are the organizational, planning, scheduling, information management, control, and project delivery subsystems. The primary responsibilities of project analysts involve ensuring the proper flow of information throughout the project system. The classical approach to the decision process follows rigid lines of organizational charts. By contrast, the systems approach considers all the interactions necessary among the various elements of an organization in the decision process.

The various elements (or subsystems) of the organization act simultaneously in a separate but interrelated fashion to achieve a common goal. This synergism helps to expedite the decision process and to enhance the effectiveness of decisions. The supporting commitments from other subsystems of the organization serve to counterbalance the weaknesses of a given subsystem. Thus, the overall effectiveness of the system is greater than the sum of the individual results from the subsystems.

The increasing complexity of organizations and projects makes the systems approach essential in today's management environment. As the number of complex projects increase, there will be an increasing need for project management professionals who can function as systems integrators. Project management techniques can be applied to the various stages of implementing a system, as shown in the following guidelines:

1. *Systems definition*: Define the system and associated problems using keywords that signify the importance of the problem to the overall organization. Locate experts in this area who are willing to contribute to the effort. Prepare and announce the development plan.
2. *Personnel assignment*: The project group and the respective tasks should be announced, a qualified project manager should be appointed, and a solid line of command should be established and enforced.
3. *Project initiation*: Arrange an organizational meeting, during which a general approach to the problem should be discussed. Prepare a specific development plan and arrange for the installation of the required hardware and tools.
4. *System prototype*: Develop a prototype system, test it, and learn more about the problem from the test results.
5. *Full system development*: Expand the prototype to a full system, evaluate the user interface structure, and incorporate user training facilities and documentation.
6. *System verification*: Involve experts and potential users, ensure that the system performs as designed, and debug the system as needed.
7. *System validation*: Ensure that the system yields the expected outputs. Validate the system by evaluating performance levels, such as the percentage of success in so many trials, measuring the level of deviation from expected outputs, and measuring the effectiveness of the system output in solving the problem.
8. *System integration*: Implement the full system as planned, ensure the system can coexist with systems already in operation, and arrange for technology transfer to other projects.
9. *System maintenance*: Arrange for the continuing maintenance of the system. Update solution procedures as new pieces of information become available. Retain responsibility for system performance or delegate to well-trained and authorized personnel.
10. *Documentation*: Prepare full documentation of the system, prepare a user's guide, and appoint a user consultant.

Systems integration permits the sharing of resources, such as physical equipment, concepts, information, and skills. Systems integration is now a major concern of many organizations. Even some of the organizations that traditionally compete and typically shun cooperative efforts are beginning to appreciate the value of integrating their operations. For these reasons, systems integration has emerged as a major interest in business. Systems integration may involve the physical integration of technical components, the objective integration of operations, the conceptual integration of management processes, or a combination of any of these.

Systems integration involves the linking of components to form subsystems and the linking of subsystems to form composite systems within a single department and/or across departments. It facilitates the coordination of technical and managerial efforts to enhance organizational functions, reduce cost, save energy, improve productivity, and increase the utilization of resources. Systems integration emphasizes the identification and coordination of the interface requirements among the components in an integrated system. The components and subsystems operate synergistically to optimize the performance of the total system. Systems integration ensures that all performance goals are satisfied with a minimum expenditure of time and resources. Integration can be achieved in several forms, including the following:

1. *Dual-use integration*: This involves the use of a single component by separate subsystems to reduce both the initial cost and the operating cost during the project life cycle.
2. *Dynamic resource integration*: This involves integrating the resource flows of two normally separate subsystems so that the resource flow from one to or through the other minimizes the total resource requirements in a project.
3. *Restructuring of functions*: This involves the restructuring of functions and the reintegration of subsystems to optimize costs when a new subsystem is introduced into the project environment.

Systems integration is particularly important when introducing a new work into an existing system. It involves coordinating new operations to coexist with existing operations. It may require the adjustment of functions to permit the sharing of resources, the development of new policies to accommodate product integration, or the realignment of managerial responsibilities. It can affect both the hardware and software components of an organization. The following are guidelines and important questions relevant for work systems integration.

- What are the unique characteristics of each component in the integrated system?
- How do the characteristics complement one another?
- What physical interfaces exist among the components?
- What data/information interfaces exist among the components?
- What ideological differences exist among the components?
- What are the data flow requirements for the components?
- Are there similar integrated systems operating elsewhere?

- What are the reporting requirements in the integrated system?
- Are there any hierarchical restrictions on the operations of the components of the integrated system?
- What internal and external factors are expected to influence the integrated system?
- How can the performance of the integrated system be measured?
- What benefit/cost documentations are required for the integrated system?
- What is the cost of designing and implementing the integrated system?
- What are the relative priorities assigned to each component of the integrated system?
- What are the strengths of the integrated system?
- What are the weaknesses of the integrated system?
- What resources are needed to keep the integrated system operating satisfactorily?
- Which section of the organization will have primary responsibility for the operation of the integrated system?
- What are the quality specifications and requirements for the integrated systems?

Work and the hierarchy of needs of workers

Maslow's *hierarchy of needs* is very much applicable in any work environment. According to Maslow's theory, the five different orders of human needs are as follows:

1. *Basic physiological needs*: Includes food, water, shelter, and the like. In modern society, the basic drives of human existence cause individuals to become involved in organizational life. People become participants in the organization that employs them. Thus, at the simplest level of human needs, people are motivated to join organizations, remain in them, and contribute to their objectives.
2. *Security and safety*: Security means many things to different people in different circumstances. For some, it means earning a higher income to ensure freedom from what might happen in case of sickness or during old age. Thus many people are motivated to work harder to seek success that is measured in terms of income. It can also be interpreted as job security. To some people such as civil servants and teachers, the assurance of life tenure and a guaranteed pension may be strong motivators in their participation in employing organizations.
3. *Social affiliation*: An employee with a reasonably well-paying and secure job will begin to feel that belonging and approval are important motivators in his or her organizational behavior.

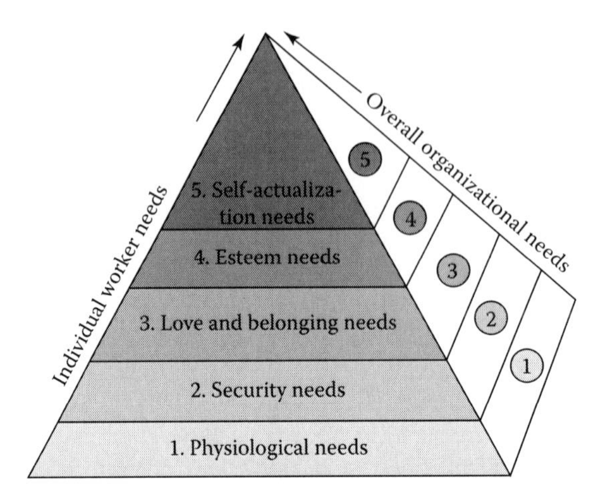

Figure 1.10 Multidimensional pyramid of the needs of workers and the organization. (Adapted from Badiru, A.B., *Triple C Model of Project Management*, CRC Press, Boca Raton, FL, 2008.)

4. *Esteem*: The need to be recognized, to be respected, and to have prestige (one's self-image and the view that one holds of oneself). There is a dynamic interplay between one's own sense of satisfaction and self-confidence on the one hand and feedback from others in such diverse forms as being asked for advice on the other hand.
5. *Self-actualization*: The desire to become more and more what one is, to become everything that one is capable of becoming. The self-actualized person is inwardly directed, seeks self-growth, and is highly motivated by loyalty to cherished values, ethics, and beliefs. Not everyone reaches the self-actualized state. It is estimated that these higher-level needs are met about 10% of the time.

In any organization, the prevailing hierarchy of the needs of the worker must be evaluated in the context of the organization's own hierarchy of needs. In this regard, Badiru (2008) developed an adaptation of the conventional triangle of hierarchy into a multidimensional pyramid of needs, as illustrated in Figure 1.10.

Work for social and economic development

Not only is work essential for personal and organizational advancement, it is also essential, from a synergistic systems perspective, for national social and economic development. The gross domestic product (GDP) is the eventual coalescing of work done at various levels of the nation. GDP is the monetary value of all the finished goods and services produced within a country in a defined period of time. Though GDP is usually calculated on an annual basis, it can be calculated on a quarterly basis as well in order to increase the granularity of management policies and practices to increase the national output. GDP includes all private and public consumption, government spending, investments, and exports (minus imports) that occur within a national boundary. In other words, GDP is a broad and composite measurement of a nation's overall economic activity. It can be calculated as follows:

$$GDP = C + G + I + NX$$

where:

C = all private consumption or consumer spending
G = the sum of government spending
I = the sum of all the country's investment, including corporate capital expenditures
NX = the nation's total net exports, calculated as total exports minus total imports

GDP is commonly used as an indicator of the economic health of a country, as well as a gauge of a country's standard of living. If a country's standard of living is high, the workers in the country do well for themselves. Since the mode of measuring GDP is uniform from country to country, GDP can be used to compare the productivity of various countries with a high degree of accuracy. Adjusting for inflation from year to year allows for an objective comparison of current GDP measurement trends. Thus, a country's GDP from any period can be measured as a percentage relative to previous years or quarters. Consequently, GDP trends can be used in measuring a nation's economic growth or decline, as well as in determining if an economy is in recession, which has a direct impact on workers and the work environment.

The concept of Total Worker Health®

Work, wellness, and wealth can go hand in hand. Workers' health directly affects GDP, based on a systems view of work, cascading from one person's level all the way (collectively) to the national level. Good health is related to good work performance. Health is an individual attribute that compliments each worker's hierarchy of needs. Without good health, even the best worker cannot perform. Without good health, even the best athlete cannot succeed. Without good health, even the most proficient expert cannot manifest his or her expertise. Without total health, even the most dedicated and experienced employee cannot contribute to the accomplishment of the organization's mission. Good health is a key part of the systems view of work advocated by this book. It is against this backdrop that the authors appreciate the concept of Total Worker Health (TWH) (Schill and

Chosewood, 2013; Schill, 2016). More details about TWH and/or pertinent worker well-being issues appear in various parts of this book. Chapter 11 presents a reprint of Schulte et al. (2015) on the considerations for incorporating well-being in the policies affecting workers and workplaces.

References

Agustiady, T. and A. B. Badiru (2012), *Sustainability: Utilizing Lean Six Sigma Techniques*, CRC Press, Boca Raton, FL.

Badiru, A. B. (2014), Quality insights: The DEJI model for quality design, evaluation, justification, and integration, *International Journal of Quality Engineering and Technology*, Vol. 4, No. 4, pp. 369–378.

Badiru, A. B. (2012), *Project Management: Systems, Principles, and Applications*, CRC Press, Boca Raton, FL.

Badiru, A. B. (2008), *Triple C Model of Project Management*, CRC Press, Boca Raton, FL.

Badiru, A. B. (2010), Half-life of learning curves for information technology project management, *International Journal of IT Project Management*, Vol. 1, No. 3, pp. 28–45.

Badiru, I. A. (2016), Comments about work management, interview with an auto industry senior engineer about corporate views of work design, Beavercreek, OH, 29 October.

GAO (Government Accountability Office) (2010), Defense Management: DOD needs better information and guidance to more effectively manage and reduce operating and support costs of major weapon systems, GAO Report No. GAO-10-717, Washington, DC, July 2010.

Geisler, J. (2012), *Work Happy: What Great Bosses Know*, Hachette, New York.

INCOSE (International Council on Systems Engineering) (n. d.), What is systems engineering? http://www.incose.org/AboutSE/WhatIsSE (accessed February 2017).

IEBOK (Industrial Engineering Body of Knowledge) (2017), published by the Institute of Industrial & Systems Engineers, Norcross, GA.

Konz, S. and S. Johnson (2004), *Work Design: Occupational Ergonomics*, 6th Edition, Holcomb Hathaway Publishers, Scottsdale, AZ.

Schill, A. L. and L. C. Chosewood (2013), The NIOSH Total Worker Health program: An overview, *Journal of Occupational and Environmental Medicine*, Vol. 55, No. 12 suppl., pp. S8–S11.

Schill, A. L. (2016), Advancing well-being through total worker health, keynote address, 17th Annual Pilot Research Project (PRP) Symposium, University of Cincinnati, Cincinnati, OH, October 13.

Schulte, P. A., R. J. Guerin, A. L. Schill, A. Bhattacharya, T. R. Cunningham, S. P. Pandalai, D. Eggerth, and C. M. Stephenson (2015), Considerations for incorporating "well-being" in public policy for workers and workplaces, *American Journal of Public Health*, Vol. 105, No. 8, pp. e31–e44, August.

Troxler, J. W. and L. Blank (1989), A comprehensive methodology for manufacturing system evaluation and comparison, *Journal of Manufacturing Systems*, Vol. 8, No. 3, pp. 176–183.

section two

Work design

chapter two

Analytics for work planning or selection

Work design can be as simple as proper work planning or selection (pick-option). This chapter presents the quantification of the *possible, implement, challenge,* and *kill* (PICK) chart for improving decisions in work design, which is the first stage of the recommended *design, evaluation, justification,* and *integration* (DEJI) model. With effective process improvement decisions and work element selection, we can improve overall organizational effectiveness, thereby leading to positive organizational transformation. Operational improvement is a goal of every organization. However, only limited quantitative approaches have been implemented specifically for that purpose. This chapter illustrates how the quantification of the PICK chart can facilitate improved operational decisions for work design or work selection. Any decision involving a work process is ultimately a decision about designing work. This connection is often made late in organizational endeavors for process improvement.

Any attempt to achieve organizational transformation must be based on leveraging effective decision-making processes within the organization. The DEJI model, by virtue of its systematic approach to efficiency, provides a strategic option for achieving the desired organizational transformation. One benefit of systems engineering is its ability to bridge the gap between quantitative and qualitative factors in the decision environment. Any decision environment will involve interaction between quantitative and qualitative information, which must be integrated for a defensible decision. For emergency and urgent decision-making needs, managers often resort to seat-of-the-pants qualitative approaches that can hardly be defended analytically, even though they possess intrinsic experiential merits. This is particularly critical in complex present-day operating environments. The popular *analytic hierarchy process* (AHP) provides a strong coupling of qualitative reasoning and quantitative analysis (Saaty, 2008; Vaidya and Kumar, 2006; Fong and Choi, 2000; Kuo, 2010). It is desirable to achieve similar quantitative and qualitative coupling for other tools, such as the PICK chart, which aids in the selection of work elements. The chart is traditionally used as a simple eyeballing device to work out package selection problems. Incorporating some element of quantification into the PICK chart will make it more defensible as an analytical tool.

The focus of this chapter is to use the quantification of the PICK chart to illustrate work package selection in a resource-constrained environment. The quantification methodology is motivated by a case study at the Air Force Institute of Technology (AFIT) (Racz et al., 2010; Badiru and Thomas, 2013). The study involved the procurement of laboratory chemicals and hazardous materials for an Environmental Safety and Occupational Health (ESOH) program. The challenge was to improve the procurement process that manages chemicals and hazardous materials for laboratories by carefully selecting task options. With effective process improvement decisions, we can improve overall organizational effectiveness, which can hopefully lead to better work designs. The need for organizational work improvement has been lamented in the literature for several years. However, only limited quantitative approaches have been implemented. The quantification technique presented here, coupled with other systems engineering tools and techniques, can facilitate enterprise process improvement and better organizational effectiveness, particularly where technology-driven learning curves are in effect (Badiru, 2012). A quantitative PICK chart approach can generate additional robust work design tools. If work design is improved at each level, it is expected that total enterprise improvement will occur.

Defense enterprise improvement case example

The military enterprise substantively and directly affects the national economy, either through direct employment, subcontracts, military construction, or technology transfer. It is thus fitting to expect that military process improvement can have a direct impact on general civilian enterprise improvement programs. Kotnour (2010, 2011) presents the fundamental elements and challenges of enterprise transformation with a view toward developing a universal framework for assessing the effectiveness of work improvement efforts. Some of the key elements suggested are as follows:

- Successful change is leadership driven.
- Successful change is strategy driven.
- Successful change is project managed.
- Successful change involves continuous learning.
- Successful change involves a systematic change process.

These elements, within the context of Air Force enterprise transformation, are all within the scope of the application of systems engineering tools and techniques. Rifkin (2011) raised questions about the time and cost elements of acquisitions in the context of enterprise transformation. Giachetti (2010) presents guidelines for designing enterprise systems for the purpose of improving decision making. These and similar references show that there is a good collection of systems engineering and business tools and techniques that the defense enterprise can adopt for work process improvement.

The pursuit of efficiencies in work programs

Functional integration and efficiencies are a primary pursuit in defense enterprises. In a report to congressional committees, the Government Accountability Office (GAO) calls for new approaches to synchronize, harmonize, and integrate the planning and operation of programs in the Intelligence, Surveillance, and Reconnaissance (ISR) enterprise of the Department of Defense (DoD) (GAO, 2008). The need for functional integration and efficiencies is depicted in Figure 2.1. The various diverse elements portrayed in the figure must be aligned and functionally integrated. Figure 2.2 is a pictorial representation of the life cycle framework for acquisitions for DoD organizations. The framework is an event-based process, where acquisition proceeds through a series of milestones associated with significant program phases. Many of these phases are amenable to the quantitative PICK charting of decisions involving work selection, cost baseline, the analysis of alternatives, resources allocation options, logistics options, and technology selection.

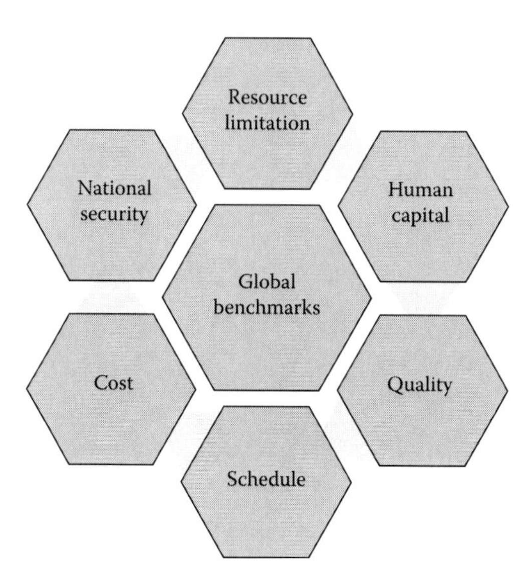

Figure 2.1 Factors of efficiencies and work integration.

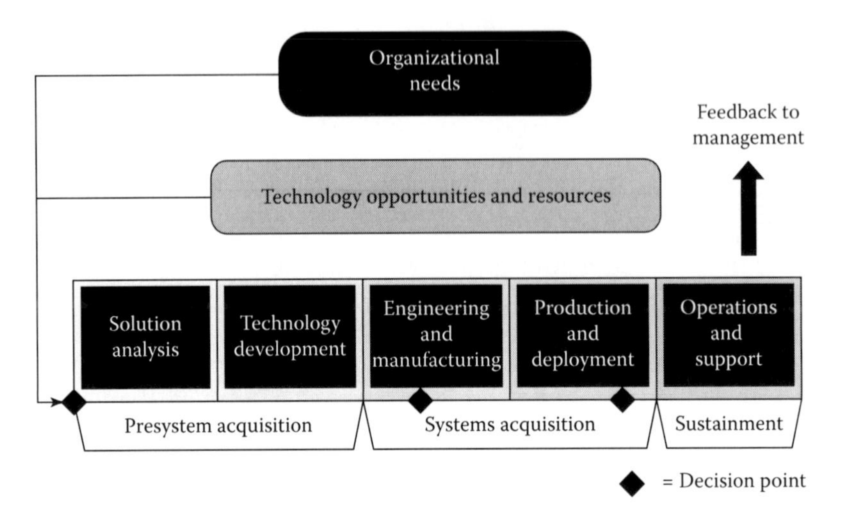

Figure 2.2 Typical organizational technology lifecycle.

In as much as DoD programs are evaluated on three primary and distinct dimensions of cost, schedule, and performance, efforts are being made within and outside the DoD to develop quantitative accountability tools for these elements. The quantification of the PICK chart fits that goal. Ward (2012) has been at the forefront of sensitizing the DoD to an integrated approach to acquisitions process improvement. With his *fast, inexpensive, simple*, and *tiny* (FIST) model, he has proposed a variety of approaches to improve cost, schedule, and performance for DoD programs. Implementing FIST in acquisitions enterprise transformation for better operational efficiency will revolve around organizational structure, process design, tools, technologies, and system architecture, all of which have embedded options and requirements. A quantitative application of the PICK chart for decisions and work selections across these elements could further enhance the concept of FIST in DoD acquisition challenges. Gibbons (2011) presents a an example case in which Starbucks instituted enterprise transformation to achieve international competitiveness. The same operational improvement that is achieved in the corporate world can be pursued in the defense enterprise. Table 2.1 compares the classical scientific management of Frederick Taylor (Taylor, 1911) with the contemporary scientific management as presently practiced by industrial and systems engineers. The taxonomy in the table can form the backdrop for the implementation of Air Force work improvement programs, such Air Force Smart Operations for the Twenty-First Century (AFSO21), as pointed out by Badiru (2007). A robust approach to using the PICK chart is desirable in cases where there is only a one-chance opportunity to make the right work selection in the decision process.

Table 2.1 Classical and contemporary principles, tools, and techniques for work management

Taylor's classical principles of scientific management	Equivalent contemporary principles, tools, and techniques	Applicability for improving acquisitions efficiency
Time studies	Work measurement; process design; PDCA; DMAIC	Effective resource allocation; schedule optimization
Functional supervision	Matrix organization structure; SMART task assignments	Team structure for efficiency
Standardization of tools and implements	Tool bins; interchangeable parts; modularity of components; ergonomics	Optimization of resource utilization
Standardization of work methods	Six Sigma processes; OODA loop	Reduction of variability
Separate planning function	Task assignment techniques; Pareto analysis	Reduction of waste and redundancy
Management by exception	Failure mode and effect analysis (FMEA); project management; Pareto analysis	Focus on vital few; task prioritization
Use of slide rules and similar time-saving devices	Blueprint templates; computer hardware and software	Use of boilerplate models
Instruction cards for workmen	Standards maps; process mapping; work breakdown structure	Reinforcement of learning
Task allocation and large bonus for successful performance	Benefit–cost analysis; value-added systems; performance appraisal	Cost reduction; productivity improvement; consistency of morale
Use of differential rate	Value engineering; work rate analysis; AHP	Input–output task coordination
Mnemonic systems for classifying products and implements	Relationship charts; group technology; charts and color coding	Goal alignment; work simplification
Routing system	Lean principles; facility layout; PICK chart; DEJI	Minimization of transportation and handling
Modern costing system	Value engineering; earned value analysis	Cost optimization

DMAIC: Define, Measure, Analyze, Improve, and Control; PDCA: Plan, Do, Check, and Act; SMART: Specific, Measurable, Achievable, Realistic, and Timed; OODA: Observe, Orient, Decide, and Act; FMEA: Failure Mode and Effect Analysis; PICK: Pick, Implement, Challenge, or Kill; DEJI: Design, Evaluate, Justify, and Integrate; AHP: Analytic Hierarchy Process.

Elements of the PICK chart

The PICK chart was originally developed by Lockheed Martin to identify and prioritize improvement opportunities in the company's process improvement applications (George, 2006). The technique is just one of the several decision tools available in process improvement endeavors. It is a very effective lean Six Sigma tool used to categorize process improvement ideas. The purpose is to qualitatively help identify the most useful ideas. A 2 × 2 grid is normally drawn on a whiteboard or large flip chart. Ideas that were written on sticky notes by team members are then placed on the grid based on a group assessment of the payoff relative to the level of difficulty. The PICK acronym comes from the labels for each of the quadrants of the grid, summarized as follows:

Possible (easy, low payoff)	Third quadrant
Implement (easy, high payoff)	Second quadrant
Challenge (hard, high payoff)	First quadrant
Kill (hard, low payoff)	Fourth quadrant

The primary purpose is to help identify the most useful ideas, especially those that can be accomplished immediately with little difficulty. These are called *just-do-its*. The general layout of the PICK chart grid is shown in Figure 2.3. The PICK process is normally done subjectively by a team of decision makers under a group decision process. This can lead to a biased and protracted debate of where each item belongs. It is desirable to improve the efficacy of the process by introducing some quantitative analysis. Just as AHP faced critics in its early years (Calantone et al., 1999; Wong and Li, 2008; Chou et al., 2004), the PICK chart is often criticized for its subjective rankings and lack of quantitative analysis. The approach presented by Badiru and Thomas (2013) alleviates such concerns by normalizing and quantifying the process of integrating the subjective rakings of those involved in the work selection of the PICK process. Human decision is inherently subjective. All we can do is to develop techniques to mitigate the subjective inputs rather than compound them with subjective summarization.

Quantitative measures of efficiency

The PICK chart may be used as a hybrid component of the existing quantitative measures of operational efficiency. Performance can be defined in terms of several organization-specific metrics. Examples are efficiency, effectiveness, and productivity, which usually go hand in hand. The existing techniques for improving efficiency, effectiveness, and productivity are quite amenable for military adaptation. Efficiency refers to the extent to which a resource (time, money, effort, etc.) is properly utilized

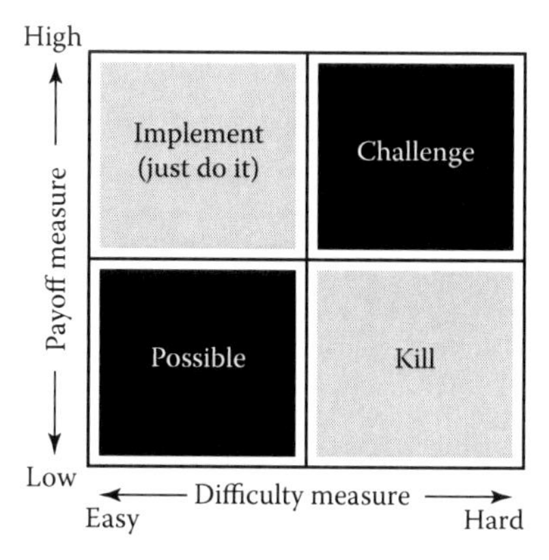

Figure 2.3 Basic layout of the PICK chart for work selection.

to achieve an expected outcome. The goal, thus, is to minimize resource expenditure, reduce waste, eliminate unnecessary effort, and maximize output. The ideal (i.e., the perfect case) is to have 100% efficiency. This is rarely possible in practice. Usually expressed as a percentage, *efficiency* (e) is computed as output over input.

$$e = \frac{\text{output}}{\text{input}} = \frac{\text{result}}{\text{effort}}$$

This ratio is also adapted for measuring productivity. For the purpose of the premise of this chapter, we offer the following definition of operational efficiency:

> *Operational efficiency* involves a scenario wherein all participants and stakeholders coordinate their respective activities, considering all the attendant factors, such that the overall organizational goals can be achieved with systematic input–process–output relationships with the minimum expenditure of resources yielding maximum possible outputs.

Effectiveness is an ambiguous evaluative term that is difficult to quantify. It is primarily concerned with achieving objectives. To model effectiveness quantitatively, we can consider the fact that an *objective* is essentially an *output* related to the numerator of the preceding efficiency equation. Thus, we can assess the extent to which the various objectives of an organization are met with respect to the available resources. Although efficiency and effectiveness often go hand in hand, they are, indeed, different and distinct. For example, one can forego efficiency for the sake of accomplishing a particular objective. Consider the statement "If we can get it done, money is no object." The military, by virtue of being mission driven, often operates this way. If, for instance, our goal is to go from point A to point B to hit a target, and we do hit the target, no matter what it takes, then we are effective. We may not be efficient based on the amount of resources expended to hit the target. For the purpose of this chapter and the use of the DEJI model, a cost-based measure of effectiveness is defined as

$$ef = \frac{s_o}{c_o}, \quad c_o > 0$$

where:

ef = the measure of effectiveness on interval (0, 1)
s_o = the level of satisfaction of the objective (rated on a scale of 0–1)
c_o = the cost of achieving the objective (expressed in pertinent cost basis: money, time, measurable resources, etc.)

If an objective is fully achieved by a work element, then its satisfaction rating will be 1. If it is not achieved at all, it will be 0. Thus, having the cost in the denominator gives a measure of achieving the objective per unit cost. If the effectiveness measures of achieving several objectives are to be compared, then the denominator (i.e., cost) will need to be normalized to a uniform scale. Overall system effectiveness can be computed as a summation as follows:

$$ef_c = \sum_{i=1}^{n} \frac{s_o}{c_o}$$

where ef_c = the composite effectiveness measure and n = the number of objectives in the effectiveness window.

Because of the potential for the effectiveness measure to be very small based on the magnitude of the cost denominator, it is essential to place it on a scale of 0–100. Thus, the highest comparative effectiveness per unit cost will be 100, while the lowest will be 0. The preceding quantitative measure of effectiveness makes most sense when comparing alternatives for achieving a specific objective. If the effectiveness of achieving an objective in absolute (noncomparative) terms is desired, it is necessary to determine the range of costs, minimum to maximum, applicable for achieving the objective. Then, we can assess how well we satisfy the objective with the expenditure of the maximum cost versus the expenditure of the minimum cost. By analogy, killing two birds with one stone is efficient. By comparison, the question of effectiveness is whether we kill a bird with one stone or kill the same bird with two stones, if the primary goal is to kill the bird nonetheless. In technical terms, systems that are designed with parallel redundancy can be effective, but not necessarily efficient. In such cases, the goal is to be effective (i.e., to get the job done) rather than to be efficient. The *productivity* of a work element is a measure of throughput per unit time. The traditional application of productivity computation is in the production environment, with countable or measurable units of output in repetitive operations. Manufacturing is a perfect scenario for productivity computations. Typical productivity formulas include the following:

$$P = \frac{Q}{q} \quad \text{or} \quad P = \frac{Q}{q}(u)$$

where:

P = productivity
Q = output quantity
q = input quantity
u = utilization percentage

Notice that Q/q also represents efficiency (i.e., output/input), as defined earlier. Applying the utilization percentage to this ratio modifies the ratio to provide actual productivity yield. For the military environment, which is a nonmanufacturing setting, productivity analysis is still of interest. The military organization is composed, primarily, of knowledge workers, whose productivity must be measured in alternate terms, perhaps through work rate analysis. Rifkin (2011) presents the following productivity equation, suitable for implementation in any work environment:

$$\text{Product (i.e., output)} = \text{productivity (objects per person-time)} \times \text{effort (person-time)}$$

where

effort = duration × the number of people.

He suggests using this measure of productivity to draw inferences about organizational transformation and work efficiency. It should be noted that higher efficiency, effectiveness, and productivity are not simply a resource availability issue. An organization with ample resources can still be inefficient, ineffective, and unproductive because of flawed work design. Thus, organizational impediments, apart from resource availability, should be identified and mitigated.

Case study of work selection process improvement

This section presents a case study of an improvement project at AFIT (Racz et al., 2010). As a part of the enterprise transformation effort of the Air Force, high-value projects are selected and targeted for the application of improvement methodologies. One such project is an acquisitions challenge in the Environmental Safety and Occupational Health (ESOH) program, in which it is desired to improve the ways AFIT procures and manages chemicals and hazardous materials for laboratories. Figure 2.4 illustrates the overall project execution environment for the improvement project. We focus on two examples of the improvement tools used during the ESOH project. The first one is a SIPOC chart, which details the integrated flow of *suppliers*, *inputs*, *process*, *outputs*, and *customers*. This is shown in Table 2.2.

The second tool of interest in this case study is a PICK chart. Figure 2.5 illustrates the PICK chart used for the ESOH project. When faced with multiple improvement ideas, a PICK chart may be used to determine the most useful one to pick. The horizontal axis, representing the ease of implementation, would typically include some assessment of the cost to implement the category. More expensive actions can be said to be more difficult to implement. Although this acquisitions example represents a simple scenario, the same tools, techniques, and decision processes used can be expanded and extended to the more complicated higher-level work selection challenges.

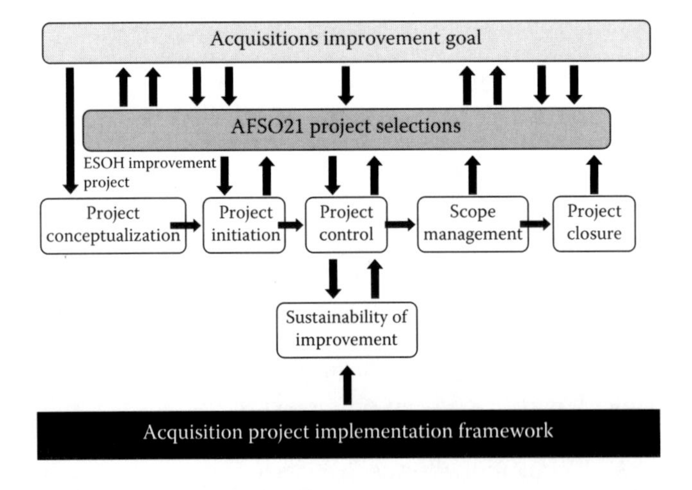

Figure 2.4 Case study: execution framework for ESOH work selection.

Table 2.2 SIPOC chart for ESOH work selection

Suppliers	Inputs	Process	Outputs	Customers
Consultants	Training	See flow charts (value stream maps)	Safe working environment	Local, state, and federal agencies; (invoices, payments)
Faculty	Purchase process			
Chemical vendors	Inventory		Compliance with Air Force, local, state, federal requirements	
Equipment vendors	Personal protective equipment		Properly trained students	Users (students, faculty, research collaborators visitors)
Base system (physical plant, chemical management, system supply/ disposal)	Site/lab survey Price quotes		Students perform excellent R&D	Research sponsors Maintenance staff
Organization management	Government purchase Card		Student education	Compliance managers
Students/ research assistant			Useable product for sponsor (equipment, publication, information)	Facility Manager
Computer Support				Air Force
Funding agencies	Institutional regulations			Institution leadership
Local business	Federal and local Law		Safety culture	
Contractors			Academic degrees	
EPA/OSHA/ Army, etc.	Time to complete forms		Contracts	
Collaboration with other colleges	Research proposal approvals		Reports to Air Force groups, contractors, etc.	
Local inventor	Equipment		Excess item disposal	
Funding source	Expertise			
Base laser safety	Sponsor requirements			
Inspectors				

SIPOC: Suppliers, Inputs, Process, Outputs, and Customers; ESOH: Environmental, Safety, and Occupational Health; EPA: Environmental Protection Agency; OSHA: Occupational Safety and Health Administration; R&D: Research and Development.

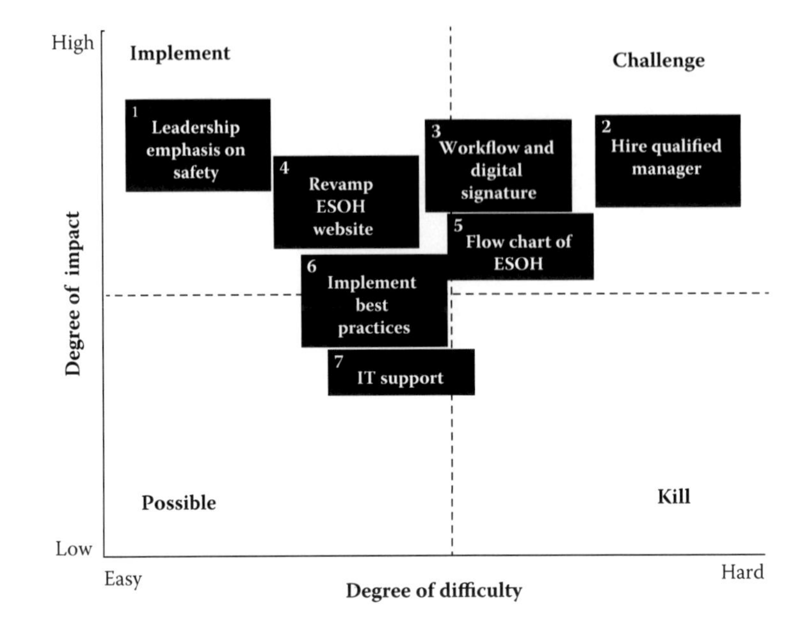

Figure 2.5 PICK chart example for ESOH improvement project.

PICK chart quantification methodology

The placement of items into one of the four categories in a PICK chart is done through expert ratings, which are often subjective and nonquantitative. In order to give some quantitative basis to the PICK chart analysis, this chapter presents the methodology of dual numeric scaling on the impact and difficulty axes. Suppose each project is ranked on a scale of 1–10 and plotted accordingly on the PICK chart. Then, each project can be evaluated on a binomial pairing of the respective rating on each scale. For our ESOH example, let x represent the level of impact and let y represent the rating along the axes of difficulty. Note that a high rating along x is desirable, while a high rating along y is undesirable. Thus, a composite rating involving x and y must account for the adverse effect of high values of y. A simple approach is to define $y' = (11 - y)$, which is then used in the composite evaluation. If there are more factors involved in the overall project selection scenario, the other factors can take on their own lettered labeling (e.g., a, b, c, etc.). Then, each project will have an n-tuple assessment vector. In its simplest form, this approach will generate a rating such as the following:

$$\text{PICK}_{R,i}(x, y') = x + y'$$

where:

$\text{PICK}_{R,i}(x,y)$ = the PICK rating of project i ($i = 1, 2, 3, \ldots, n$)
n = the number of the project under consideration
x = the rating along the impact axis ($1 \le x \le 10$)
y = the rating along the difficulty axis ($1 \le y \le 10$)
y' = $(11 - y)$

If $x + y'$ is the evaluative basis, then each project's composite rating will range from 2 to 20, 2 being the minimum and 20 being the maximum possible. If $(x)(y)$ is the evaluative basis, then each project's composite rating will range from 1 to 100. In general, any desired functional form may be adopted for the composite evaluation. Another possible functional form is as follows:

$$\begin{aligned}\text{PICK}_{R,i}(x, y'') &= f(x, y'') \\ &= (x + y'')^2\end{aligned}$$

where y'' accounts for the converse impact of the axes of difficulty. This methodology provides a quantitative measure for translating the entries in a conventional PICK chart into an analytical ranking of the improvement alternatives, thereby reducing the level of subjectivity in the final decision. The methodology can be extended to cover cases where a project has the potential to create negative impacts that will impede organizational advancement. Referring back to the PICK chart for our ESOH example, we develop the numeric illustration shown in Table 2.3.

As expected, the highest $x + y'$ composite rating (i.e., 18) is in the second quadrant, which represents the *implement* region. The lowest composite rating is 10 in the first quadrant, which is the *challenge* region. With this type of quantitative analysis, it becomes easier to design, evaluate, justify, and integrate (i.e., apply the DEJI model). This facilitates a more

Table 2.3 Numeric evaluation of PICK chart rating for ESOH work elements

Improvement project	x *Rating*	y *Rating*	$y' = 11 - y$	$x + y'$	xy'
1. Leadership emphasis	9	2	9	**18**	81
2. Full-time issue manager	9	10	1	**10**	9
3. Work flow digital signature	9	6	5	14	45
4. Work group process	8	3	8	16	64
5. Work flow chart VSM	7	6	5	12	35
6. Implement best practices	7	4	7	14	49
7. Support center other	6	4	7	13	42

rigorous analytical technique compared with the traditional subjective arm-waving approaches. One concern is that although quantifying the placement of alternatives on the PICK chart may improve the granularity of the relative locations on the chart, it still does not eliminate the subjectivity of how the alternatives are assigned to quadrants in the first place. This is a recognized feature of many decision tools and can be mitigated by the use of additional techniques that aid decision makers to refine their choices. AHP could be useful for this purpose. Quantifying subjectivity is a continuing challenge in decision analysis. The PICK chart quantification approach offers an improvement over the conventional approach.

Implementing the PICK chart

Although the PICK chart has been used extensively in industry, there are few published examples in the open literature. The tool is effective for managing process enhancement ideas and classifying them during the *identify and prioritize opportunity* phase of a Six Sigma project. When a process improvement team is faced with multiple improvement ideas, the PICK chart helps address issues related to deciding which ideas should be implemented and which work elements should be embraced. The steps for implementing a PICK chart are as follows:

Step 1: Place the subject question on a chart. The question needs to be asked and answered by the team at different stages to be sure that the data that are collected are relevant.

Step 2: Put each component of the data on a different note, such as a Post-its or a small card. These notes should be arranged on the left-hand side of the chart.

Step 3: Each team member must read all notes individually and consider their importance. The team member should decide whether the element should or should not remain a fraction of the significant sample. The notes are then removed and moved to the other side of the chart. Now, the data are condensed enough to be processed for a particular purpose by means of tools such as *KJ analysis*, which is a group-focusing approach developed by the Japanese Jiro Kawakita to quickly allow groups to reach a consensus on priorities of subjective and qualitative data.

Step 4: Apply the quantification methodology previously presented to normalize the qualitative inputs of the team.

Summary

Human uncertainty and personal preferences often creep into corporate decision processes. Incorporating some quantifiable measure is a good way to mitigate the adverse effects of qualitative reason. The quantification of the PICK chart fits the systematic approach of the DEJI model.

References

Badiru, A. B. (2007), Air Force Smart Operations for the 21st Century, *OR/MS Today*, Vol. 34, No. 1, p. 28.

Badiru, A. B. (2012), Half-life learning curves in the defense acquisition lifecycle, *Defense Acquisition Research Journal*, Vol. 19, No. 3, pp. 283–308.

Badiru, A. B. and M. Thomas (2013), Quantification of the PICK chart for process improvement decisions, *Journal of Enterprise Transformation*, Vol. 3, No. 1, pp. 1–15.

Calantone, R. J., C. A. Di Benedetto, and J. B. Schmidt (1999), Using the analytic hierarchy process in new product screening, *Journal of Product Innovation Management*, Vol. 16, pp. 65–76.

Chou, Y., C. Lee, and J. Chung (2004), Understanding m-commerce payment systems through the analytic hierarchy process, *Journal of Business Research*, Vol. 57, No. 12, pp. 1423–1430.

Fong, P. S. and S. K. Choi (2000), Final contractor selection using the analytic hierarchy process, *Construction Management and Economics*, Vol. 18, No. 5, pp. 547–557.

GAO (Government Accountability Office) (2008), Intelligence, surveillance, and reconnaissance: DOD can better assess and integrate ISR capabilities and oversee development of future ISR requirements, GAO Report No. GAO-08-374, Washington, DC, March 2008.

George, M. L. (2006), *Lean Six Sigma for Services*, McGraw-Hill, Seoul, Korea.

Giachetti, R. E. (2010), *Design of Enterprise Systems: Theory, Architecture, and Methods*, CRC Press, Boca Raton, FL.

Gibbons, P. (2011), Notes from the field: Transforming the Starbucks experience, *Journal of Enterprise Transformation*, Vol. 1, No. 1, pp. 7–13.

Kotnour, T. (2011), An emerging theory of enterprise transformations, *Journal of Enterprise Transformation*, Vol. 1, No. 1, pp. 48–70.

Kotnour, T. G. (2010), *Transforming Organizations: Strategies and Methods*, CRC Press, Boca Raton, FL.

Kuo, T. C. (2010), Combination of case-based reasoning and analytical hierarchy process for providing intelligent decision support for product recycling strategies, *Expert Systems with Applications: An International Journal*, Vol. 37, No. 8, pp. 5558–5563.

Racz, L. et al. (2010), AFIT Environmental Safety and Occupational Health (ESOH) AFSO21 event, report of the AFSO21 improvement project, AFIT, December 2010.

Rifkin, S. (2011), Raising questions: How long does it take, how much does it cost, and what will we have when we are done? What we do not know about enterprise transformation, *Journal of Enterprise Transformation*, Vol. 1, No. 1, pp. 34–47.

Saaty, T. L. (2008), Decision making with the analytic hierarchy process, *International Journal of Services Sciences*, Vol. 1, No. 1, pp. 83–98.

Taylor, F. W. (1911), *The Principles of Scientific Management*, Harper Bros, New York.

Vaidya, O. S. and S. Kumar (2006), Analytic hierarchy process: An overview of applications, *European Journal of Operational Research*, Vol. 169, No. 1, pp. 1–29.

Ward, D. (2012), Faster, better, cheaper: Why not pick all three?, *National Defense Magazine*, pp. 1–2.

Wong, J. K. W. and H. Li (2008), Application of the analytic hierarchy process (AHP) in multi-criteria analysis of the selection of intelligent building systems, *Building and Environment*, Vol. 43, No. 1, pp. 108–125.

chapter three

Learning curve analysis for work design

The illiterate of the 21st century will not be those who cannot read and write, but those who cannot learn, unlearn, and relearn.

Alvin Toffler

Whenever one person is found adequate to the discharge of a duty by close application thereto, it is worse executed by two persons, and scarcely done at all if three or more are employed therein.

George Washington

One vital basis for work performance is the impact of learning. The opening quotes highlight the importance of learning in the design and execution of work. When a work element has a certain labor component and repetition of the same work element makes the labor requirement more proficient and efficient, the work is completed more quickly, thus reducing the labor cost while preserving or increasing the level of quality of the output. Consequently, the impact of learning has the following benefits:

- Lower cost of work at the individual worker level
- Higher quality of work output
- Better customer satisfaction
- Higher productivity
- Higher efficiency
- Better organizational throughput
- Smoother coordination of work
- Less stress on the worker
- Lower production cost at the organizational level

These benefits have been modeled mathematically in various forms for decades, ever since Wright (1936) first presented the impact of learning curves on the production cost of airplanes. Grahame (2010) presents the tradition of applying learning curves to the production of not only aircraft but also spacecraft. The impact of learning can be subtle in many cases. So, analysis must rigorously track and document the durations of work elements over time as repeated production progresses. For this reason, a learning curve is sometimes referred to as a *manufacturing progress function*. This is a major topic of interest in operations analysis and improvement in industry, where work measurement provides the basis for productivity improvement.

Industrial innovation and economic advancement owe their sustainable foundations to work measurement processes (Badiru, 2008). We must measure work accurately in order to improve it. Such pursuit of work measurement accuracy requires an assessment of the impact of learning during the production cycle. This chapter presents the basic foundations for modeling and utilizing learning curves. Because of the multitude of factors influencing work performance, the analytic framework for learning curves often extend from univariate to multivariate models. Multivariate learning curve functions are essential to accounting for the variety of factors that can influence production operations. In many industrial operations, subtle, observable, quantitative, and qualitative factors intermingle to compound the productivity analysis problem. Such factors are best modeled within a multivariate learning curve framework. This chapter presents learning curve examples to illustrate the modeling process. The chapter also includes computational procedures for work rate analysis.

Learning curves for better work design

The opportunity to work translates into the opportunity to learn and improve with specific work elements. But we must measure work before we can improve it as part of the evaluation of human performance (Wilson and Corlett, 2005). This conveys the importance of work measurement in industry, business, and government. What is not measured cannot be improved. Thus, work measurement should be a key strategy for productivity measurement. The following quote provides an inspirational foundation for the pursuit of work measurement activities from the standpoints of both research and application.

> No man can efficiently direct work about which he knows nothing.
>
> **Thurman H. Bane**

Work measurement for better work design

It is when the ramifications of learning are understood through work measurement that we can have a better basis for designing and redesigning work. Aft (2010) suggests that the academic research of work measurement has real-world value and practical applications for industry. He argues that standards obtained through work measurement provide essential information for the success of an organization. Such pieces of information include the following:

- Data for scheduling
- Data for staffing
- Data for line balancing
- Data for materials requirement planning
- Data for wage payments
- Data for costing
- Data for employee evaluation

In each of these data types, the effect of learning in the presence of multiple factors should be taken into consideration for operational decision-making purposes.

Foundation for learning curve analysis

A learning curve, in the context of work design, refers to the improved productivity obtained from the repetition of work elements. Several research studies have confirmed that human performance improves with reinforcement or frequent repetition (Carr, 1946; Conley, 1970; Hirchmann, 1964; Belkaoui, 1986; Yelle, 1979). Reduced operation processing times achieved through learning curve effects can directly translate to cost savings for manufacturers and improved morale for employees (Badiru, 1988). Learning curves are essential for setting production goals, monitoring progress, reducing waste, and improving efficiency (Knecht, 1974; Richardson, 1978; Conway and Schultz, 1959; Steven, 1999). The applications of learning curves extend well beyond conventional productivity analysis. For example, Dada and Srikanth (1990) examine the impact of accumulated production on monopoly pricing decisions.

Typical learning curves show the relationship between cumulative average production cost per unit and cumulative production volume based on the effect of learning. For example, an early study by Wright (1936) disclosed an 80% learning effect, which indicates that a given operation is subject to a 20% productivity improvement each time the production quantity doubles. This productivity improvement phenomenon is illustrated later in the chapter. With information about expected future productivity levels, a learning curve can serve as a predictive tool for obtaining time estimates for tasks that are repeated within a production cycle (Chase and Aquilano, 1981) or for project management (Teplitz and Amor, 1998). In any work environment, tangible, intangible, quantitative, and qualitative factors intermingle to compound the output of work. Consequently, a more comprehensive analytical methodology and human-centric considerations are essential for better work design.

Some of the specific analyses that can benefit from the results of multifactor learning curves include cost estimation, work simplification, break-even analysis, manpower scheduling, make-or-buy decisions, production planning, budgeting and resource allocation, and productivity management.

Univariate learning curve models

The conventional univariate learning curve model presents several limitations in practice. Since the first formal publication of learning curve theory by Wright (1936), there have been numerous alternative propositions concerning the geometry and functional forms of learning curves (Baloff, 1971; Jewell, 1984; Kopcso, 1983; Smunt, 1986; Towill and Kaloo, 1978; Yelle, 1983). Some of the classical models include the log-linear model, the S-curve, the Stanford-B model, DeJong's learning formula, Levy's adaptation function, Glover's learning formula, Pegels's exponential function, Knecht's upturn model, Yelle's product model, and the multiplicative power model (Asher, 1956; DeJong, 1957; Levy, 1965; Glover, 1966; Pegels, 1969; Knecht, 1974; Yelle, 1976; Waller and Dwyer, 1981).

Log-linear learning curve model

The log-linear model is by far the most common model of learning curves. The model states that the improvement in productivity is constant (i.e., it has a constant slope) as output increases. The basic log-linear model is shown in Figure 3.1.

There are two forms of the log-linear model: the average cost model and the unit cost model, the former of which is used most often. It specifies the relationship between the cumulative average cost per unit and

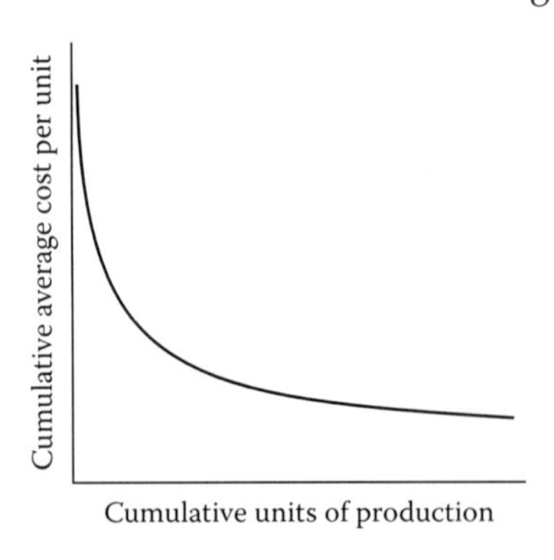

Figure 3.1 Basic log-linear learning curve model.

cumulative production. The relationship indicates that the cumulative cost per unit will decrease by a constant percentage as the cumulative production volume doubles. The model is expressed as

$$A_x = C_1 x^b$$

where:

A_x = the cumulative average cost of producing x units
C_1 = the cost of the first unit
x = the cumulative production count
b = the learning curve exponent (i.e., constant slope of the curve on a log-log paper)

On a log-log paper, the log-linear model is represented as a straight line, as follows:

$$\log A_x = \log C_1 + b \log x$$

It is from this straight-line equation that the name *log-linear* model is derived.

Example

Assume that 50 units of an item are produced at a cumulative average cost of $20 per unit. Suppose we want to compute the learning percentage when 100 units are produced at a cumulative average cost of $15 per unit. The learning curve analysis would proceed as follows:

Initial production level = 50 units; average cost = $20
Double production level = 100 units; cumulative average cost = $15

Using the log relationship, we obtain the following equations:

$$\log 20 = \log C_1 + b \log 50$$

$$\log 15 = \log C_1 + b \log 100$$

Solving the equations simultaneously yields

$$b = \frac{\log 20 - \log 15}{\log 50 - \log 100} = -0.415$$

Thus

$$p = (2)^{-0.415} = 0.75$$

That is, there is a 75% learning rate. In general, the learning curve exponent, b, may be calculated directly from actual data or computed analytically, as follows:

$$b = \frac{\log A_{x1} - \log A_{x2}}{\log x_1 - \log x_2}$$

$$b = \frac{\ln(p)}{\ln(2)}$$

where:

x_1 = the first production level
x_2 = the second production level
A_{x1} = the cumulative average cost per unit at the first production level
A_{x2} = the cumulative average cost per unit at the second production level
p = the learning rate percentage

Using the basic cumulative average cost function, the total cost of producing x units is computed as

$$TC_x = (x)A_x = (x)C_1x^b = C_1x^{(b+1)}$$

The unit cost of producing the xth unit is given by

$$U_x = C_1x^{(b+1)} - C_1(x-1)^{(b+1)}$$
$$= C_1x\left[x^{(b+1)} - (x-1)^{(b+1)}\right]$$

The marginal cost of producing the xth unit is given by

$$MC_x = \frac{d[TC_x]}{dx} = (b+1)C_1x^b$$

Example

Suppose in a production run of a certain product it is observed that the cumulative hours required to produce 100 units is 100,000 hours with a learning curve effect of 85%. For project planning purposes, an analyst needs to calculate the number of hours spent in producing the 50th unit. Following the notation used previously, we gather the following information:

$$p = 0.85$$

$$X = 100 \text{ units}$$

$$A_x = 100{,}000 \text{ hours}/100 \text{ units} = 1{,}000 \text{ hours/unit}$$

Now

$$0.85 = 2^b$$

Therefore,

$$b = -0.2345$$

Also,

$$100,000 = C_1(100)^b$$

Therefore, $C_1 = 2944.42$ hours. Thus,

$$C_{50} = C_1(50)^b = 1176.50 \text{ hours}$$

That is, the cumulative average hours for 50 units is 1176.50 hours. Therefore, cumulative total hours for 50 units = 58,824.91 hours. Similarly,

$$C_{49} = C_1(49)^b = 1182.09 \text{ hours}$$

That is, the cumulative average hours for 49 units is 1182.09 hours. Therefore, cumulative total hours for 49 units = 57,922.17 hours. Consequently, the number of hours for the 50th unit is given by

$$58,824.91 \text{ hours} - 57,922.17 \text{ hours} = 902.74 \text{ hours}$$

The unit cost model is expressed in terms of the specific cost of producing the xth unit. The unit cost formula specifies that the individual cost per unit will decrease by a constant percentage as cumulative production doubles. The formulation of the unit cost model is as follows: Define the average cost as $A_x = C_1x^b$, so that the total cost is $TC_x = (x)A_x = (x)C_1x^b = C_1x^{(b+1)}$ and the marginal cost is given by $MC_x = d[TC_x]/dx = (b+1)C_1x^b$. This is the cost of one specific unit. Therefore, we define the marginal unit cost model as $U_x = (1+b)C_1x^b$, which is the cost of producing the xth unit.

The following formulas are derived from the preceding equations and shows the relationship between A_x and U_x.

$$U_x = (1+b)A_x$$

$$C_1x^b = \frac{U_x}{(1+b)}$$

$$C_1 = \frac{U_x x^{-b}}{(1+b)}$$

Multifactor learning curves

Extensions of the single-factor learning curve are important for the realistic analysis of productivity gain. In project operations, several factors can intermingle to affect performance. Heuristic decision making, in particular, requires careful consideration of qualitative factors. There are numerous factors that can influence how fast, how far, and how well a worker learns within a given time span. Multivariate models are useful for performance analysis in project planning and control. One form of the multivariate learning curve is defined as

$$A_x = K\prod_{i=1}^{n} c_i x_i^{b_i}$$

where:

A_x = the cumulative average cost per unit
K = the cost of the first unit of the product
x = the vector of specific values of independent variables
x_i = the specific value of the ith factor
n = the number of factors in the model
c_i = the coefficient for the ith factor
b_i = the learning exponent for the ith factor

A bivariate form of the model is presented as follows:

$$C = \beta_0 x_1^{\beta_1} x_2^{\beta_2}$$

where C is a measure of cost, and x_1 and x_2 are independent variables.

The bivariate model is used here to illustrate the natural and modeling approach for general multivariate models. An experiment conducted by the author models a learning curve containing two independent variables: *cumulative production* (x_1) and *cumulative training time* (x_2). The following model was chosen for illustration purposes.

$$A_{x_1 x_2} = K c_1 x_1^{b_1} c_2 x_2^{b_2}$$

where:

A_x = the cumulative average cost per unit for a given set X of factor values
K = the intrinsic constant
x_1 = the specific value of the first factor
x_2 = the specific value of the second factor
c_i = the coefficient for the ith factor
b_i = the learning exponent for the ith factor

Sample production data were used to develop a computational model. In the data set, two data replicates are used for each of the 10 combinations of cost and time values. Observations are recorded for the number of units representing double production volumes. The model was transformed into the following natural logarithmic form:

$$\ln A_x = \left[\ln K + \ln\left(c_1c_2\right)\right] + b_1 \ln x_1 + b_2 \ln x_2 = \ln a + b_1 \ln x_1 + b_2 \ln x_2$$

where a represents the combined constant in the model, such that $a = (K)(c_1)(c_2)$. A regression approach yielded the following fitted model:

$$\ln A_x = 5.70 - 0.21(\ln x_1) - 0.13(\ln x_2)$$

$$A_x = 298.88 x_1^{-0.21} x_2^{-0.13}$$

with an R^2 value of 96.7%. In this model, $\ln(a) = 5.70$ (i.e., $a = 298.88$), x_1 = the cumulative production units, and x_2 = the cumulative training time in hours. As in the univariate case, the bivariate model indicates that the cumulative average cost decreases as cumulative production and training time increase. For a production level of 1750 units and a cumulative training time of 600 hours, the fitted model indicates an estimated cumulative average cost per unit as follows:

$$A_{(1750,600)} = (298.88)\left(1750^{-0.21}\right)\left(600^{-0.13}\right) = 27.12$$

Similarly, a production level of 3500 units and training time of 950 hours yield the following cumulative average cost per unit:

$$A_{(3500,950)} = (298.88)\left(3500^{-0.21}\right)\left(950^{-0.13}\right) = 22.08$$

To use the fitted model, consider the following problem: The standards department of a manufacturing plant has set a target average cost per unit of $12.75 to be achieved after 1000 hours of training. We want to find the cumulative units that must be produced in order to achieve the target cost. From the fitted model, the following expression is obtained.

$$\$12.75 = (298.88)\left(X^{-0.21}\right)\left(1{,}000^{-0.13}\right)$$

$$X = 46{,}409.25$$

On the basis of the large number of cumulative production units required to achieve the expected standard cost, the standards department may want to review the cost standard. The standard of $12.75 may not be achievable if there is limited market demand (i.e., demand is much less than 46,409 units) for the particular product being considered. The relatively flat surface of the learning curve model as units and training time increase implies that more units will need to be produced in order to achieve any additional significant cost improvements. Thus, even though an average cost of $22.08 can be obtained at a cumulative production level of 3500 units, it takes several thousand additional units to bring the average cost down to $12.75 per unit.

Interruption of learning

Interrupting the learning process can adversely affect expected performance and work may need to be redesigned. An example of how to address this is the *manufacturing interruption ratio,* which considers the learning decay that occurs when a learning process is interrupted. One possible expression for the ratio is as follows:

$$Z = (C_1 - A_x)\frac{(t-1)}{11}$$

where:

Z = the per-product loss of learning costs due to manufacturing interruption
t = 1, 2, …, 11 (months of interruption from 1–11 months)
C_1 = the cost of the first unit of the product
A_x = the cost of the last unit produced before a production interruption

The unit cost of the first unit produced after production begins again is given by

$$\begin{aligned} A_{(x+1)} &= A_x + Z \\ &= A_x + (C_1 - A_x)\frac{(t-1)}{11} \end{aligned}$$

Interruption to the learning process can be modeled by incorporating forgetting functions into regular learning curves. In any practical situation, an allowance must be made for the potential impacts that forgetting may have on performance. Potential applications of the combined learning and forgetting models include designing training programs, manufacturing economic analysis, manpower scheduling, production planning, labor estimation, budgeting, and resource allocation.

Learning curves in worker health care

As in the production environment, earning curves are applicable to the routine and repetitive aspects of surgical procedures (Ernst and Szczesny,

2005). Each surgery is unique and may involve things that have not been seen before. Thus, learning curve analysis must be adapted to this unique application. Even a doctor who has performed a particular surgery many, many times over several years may still have to deal with a new and unique procedure. The best way to apply learning curves for performance improvement in surgical operations is to first determine those functions that are routine or standardized, those in which we can expect to make learning curve improvements. There still needs to be an allowance for procedures that are new or unique and may require more care and time on the part of the surgeon. In other words, there will be some incompressible functions in the surgical process. As in the case of unplanned production interruption, which is one of the things that may be experienced during surgery in a hospital, learning and forgetting do occur in health care delivery projects (i.e., operations).

The theory of learning curves states that direct labor improves as it works longer and performs more repetitions. The more often a worker repeats a given task, the more efficient he or she will become. This phenomenon of learning also occurs for a doctor who performs a surgery through repetition in the health care industry. During the process of operation scheduling, a hospital may need to assign the duration for every surgery in an operating room. It is known that different kinds of surgeries require different durations. Even for the same kind of surgery, if it is carried out by different doctors, the durations will also be different. Besides, even for the same doctor performing the same kind of surgery, the durations can still be significantly different, depending on the prevailing circumstances and needs. Hospital administrators often find out that there is a lot of wasted time during surgical operations. If they hope to improve the efficiency of the surgical staff (surgeons, nurses, etc.), learning curves could be one useful approach. The goal is to find an analytical approach to determine a "standard duration" for a doctor to perform a certain kind of surgery. But questions to address include the following:

1. The theory and models of learning curves have been widely used in low-skill and high-volume repetitive manufacture. Is the conventional production-based theory applicable to high-skill and one-of-a-kind operations in the operating room?
2. There are incompressible task durations in the operating room, regardless of the number of repetitions. Which learning curve models are relevant for predicting the duration of surgery?

As reiterated by Jaber and Guiffrida (2008), learning curves are applicable to surgery whether it is directly performed by a surgeon (hand–body contact) or through a medium (telesurgery). The learning curves for both types of surgery are different. The time to complete surgery reduces with time, but not always. Surgery involves a set of sequential steps where a surgeon allows some time for feedback (reaction time) from the patient. If there are complications—that is, the patient's body reacts negatively to the procedure—then a corrective action must be performed by the surgeon to alleviate any risk to the patient. Such complications can extend the time of a surgery; however, it does not necessarily mean that there is no learning on the part of the surgeon. These complications may bring learning opportunities.

Each type of surgery may follow a different learning curve. Also, if there is a knowledge-sharing mechanism among the surgeons (i.e., supporting complementary mutual experiences or an experience repository bank in a hospital), then they will always progress on their respective learning curves.

Wide applications of learning curves

A multivariate learning curve model extends the conventional analysis of the effect of learning to become a more robust decision-making tool for operations management. As pointed out by Smith (1989), learning curves can be used to plan manpower needs; set labor standards; establish prices; analyze, make, or buy decisions; review wage incentive payments; evaluate organizational efficiency; develop quantity sales discounts; evaluate employee training needs; evaluate capital equipment alternatives; predict future production unit costs; create production delivery schedules; and develop work design strategies. The parameters that affect the learning rate are broad in scope. Unfortunately, practitioners do not fully comprehend the causes of and influences on learning rates. Learning is typically influenced by many factors in a manufacturing operation. Such factors may include

Governmental factors

- Industry standards
- Educational facilities
- Regulations
- Economic viability
- Social programs
- Employment support services
- Financial and trade supports

Organizational factors

- Management awareness
- Training programs
- Incentive programs
- Employee turnover rate

- Cost accounting system
- Labor standards
- Quality objectives
- Pool of employees
- Departmental interfaces
- Market and competition pressures
- Obsolescence of machinery and equipment
- Long-range versus short-range goals

Product-based factors

- Product specs
- Raw material characteristics
- Environmental impacts of production
- Delivery schedules
- Number and types of constituent operations
- Product diversity
- Assembly requirements
- Design maturity
- Work design and simplification
- Level of automation

Multivariate learning curve models offer a robust tool through which several of the aforementioned factors may be quantitatively evaluated simultaneously. Decisions made on the basis of multivariate analyses are generally more reliable than decisions based on single-factor analyses. The following sections present brief outlines of specific examples of the potential applications of multivariate learning curve models.

Designing training programs

As shown in the bivariate model, the need for employee training can be assessed with the aid of a multivariate learning curve model. Projected productivity gains are more likely to be met if more of the factors influencing productivity can be included in the conventional analysis.

Manufacturing economic analysis

The results of multivariate learning curve analysis are important for various types of economic analysis in the manufacturing environment. The declining state of manufacturing in many countries has been a subject of much discussion in recent years. A reliable methodology for the cost analysis of manufacturing technology for specific operations is essential to the full exploitation of the recent advances in the available technology.

Manufacturing economic analysis is the process of evaluating manufacturing operations on a cost basis. In manufacturing systems, many tangible, intangible, quantitative, and qualitative factors intermingle to compound the cost analysis problem. Consequently, a more comprehensive evaluation methodology, such as a multivariate learning curve, can be very useful.

Break-even analysis

The conventional break-even analysis assumes that the variable cost per unit is a fixed value. On the contrary, learning curve analysis recognizes the potential reduction in variable cost per unit due to the effect of learning. Due to the multiple factors involved in manufacturing, multivariate learning curve models should be investigated and adopted for break-even cost analysis.

Make-or-buy decisions

Make-or-buy decisions can be enhanced by considering the effect of learning on items that are manufactured in-house. Make-or-buy analysis involves a choice between the cost of producing an item and the cost of purchasing it. Multivariate learning curves can provide the data for determining the accurate cost of producing an item. A make-or-buy analysis can be coupled with a break-even analysis to determine when to make or buy a product.

Manpower scheduling

A consideration of the effect of learning in the manufacturing environment can lead to a more accurate analysis of manpower requirements and the accompanying schedules. In integrated production, where parts move sequentially from one production station to another, the effect of multivariate learning curves can become even more applicable. The allocation of resources during production scheduling should not be made without considering the effect of learning (Liao 1979).

Production planning

The overall production planning process can benefit from multivariate learning curve models. Preproduction planning analysis of the effect of multivariate learning curves can identify areas where better and more detailed planning may be needed. The more preproduction planning that is done, the better the potential for achieving the productivity gains that are produced by the effects of learning.

Labor estimating

Carlson (1973) showed that the validity of log-linear learning curves may be suspect in many labor analysis problems. For manufacturing activities involving operations in different stations, several factors interact to determine the learning rate of workers. Multivariate curves can be of use in developing accurate labor standards in such cases. Multivariate learning curve analysis can complement conventional work measurement studies.

Budgeting and resource allocation

Budgeting or capital rationing is a significant effort in any manufacturing operation. Multivariate learning curve analysis can provide a management guide for allocating resources to production operations on a more equitable basis. The effects of learning can be particularly useful in zero-base budgeting policies (Badiru, 1988). Other manufacturing cost analyses where a multivariate learning curve analysis could be of use include bidding (Yelle, 1979), inventory analysis, productivity improvement programs (Towill and Kaloo, 1978), goal setting (Richardson, 1978), and lot sizing (Kopcso, 1983).

Takt time for work planning

Activity planning and work scheduling are not totally unlike production planning functions. Thus, general work planning can benefit from techniques typically used in the production environment. One such technique is *takt time* computation. *Takt* is the German word referring to how an orchestra conductor regulates the speed, beat, or timing so that the orchestra plays in unison. So, the idea of Takt time is to regulate the rate time or pace of producing a completed product. This refers to the production pace at which workstations must operate in order to meet a target production output rate. In other words, it is the pace of production needed to meet customer demand. The production output rate is set based on product demand. In a simple sense, if 2000 units of a widget are to be produced within an 8-hour shift to meet a market demand, then 250 units must be produced per hour. That means, a unit must be produced every $60/250 = 0.24$ minutes (14.8 seconds). Thus, the Takt time is 14.4 seconds. Lean production planning then requires that workstations be balanced such the production line generates a product every 14.4 seconds. This is distinguished from the cycle time, which refers to the actual time required to accomplish each workstation task. Cycle time may be less than, more than, or equal to takt time. Takt is not a number that can be measured with a stop watch. It must be calculated based on the prevailing production needs and scenario. Takt time equation is:

$$T = \frac{T_a}{T_d}$$

$$= \frac{\text{available work time} - \text{breaks}}{\text{customer demand}}$$

$$= \frac{\text{net available time per day}}{\text{customer demand per day}}$$

where:

T = takt time (in minutes of work per unit produced)
T_a = net time available to work (in minutes of work per day)
T_d = time demand (i.e., customer demand in units required per day)

Takt time is often expressed as *seconds per piece*, indicating that customers are buying a product once every so many seconds. Takt time is not expressed as *pieces per second*.

The objective of lean production is to bring the cycle time as close to the takt time as possible; that is choreographed. In a balanced line design, the takt time is the reciprocal of the production rate.

Improper recognition of the role of takt time can make an analyst to overestimate the production rate capability of a line. Many manufacturers have been known to overcommit to customer deliveries without accounting for the limitations imposed by takt time. Since takt time is set based on customer demand, its setting may lead to an unrealistic expectations of workstations. For example, if the constraints of the prevailing learning curve will not allow sufficient learning time for new operators, then takt times cannot be sustained. This may lead to the need for buffers to temporarily accumulate units at some workstations. But this defeats the pursuits of lean production or just-in-time. The need for buffers is a symptom of imbalances in takt time. Some manufacturers build *takt gap* into their production planning for the purpose of absorbing nonstandard occurrences in the production line. However, if there are more nonstandard or random events than have been planned for, then production rate disruption will occur.

It is important to recognize that the maximum production rate determines the minimum takt time for a production line. When demand increases, takt time should be decreased. When demand decreases, takt time should be increased. Production crew size plays an important role in setting and meeting takt time. The equation for calculating the crew size for an assembly line doing one piece flow that is paced to takt time is as follows:

$$\text{Crew size} = \frac{\text{sum of manual cycle times}}{\text{takt time}}$$

Even though takt time is normally used in production operations, the concept and computations are applicable to resource allocation and usage in project management. It can be adapted for coordination project task completion rates. In that case, *customer demand* can be defined in terms of work flow requirements to keep the project schedule on track in accordance with the desired project completion target dates or volume of output. For example, if a construction project calls for laying 500 blocks within a work day of 7.5 hours, then takt time would be calculated as follows:

$$T = \frac{7.5 \text{ hours}}{500 \text{ blocks}}$$

$$= 0.015 \text{ hours/blocks}$$

$$= 0.9 \text{ minutes/block (i.e., 54 seconds per block)}$$

Depending on the experience of the masons and the desired quality of the output (excessive pace degrades quality), this pace may be judged unachievable. Thus, necessitating an adjustment of the goal, modifying the work schedule, or allocating more resources.

Using takt time for work design

The following guidelines are useful for using the concept of takt time for work design.

- Carefully evaluate the demand for the work (i.e., what does the downstream work flow really need?). *Downstream* refers to an activity sequence that occurs later on in a project schedule.
- Accurately assess the available work time. In addition to approved breaks, consider other "time robbers" that workers experience during the work day, either as an imposed reality or as self-initiated distractions. Not all such "illegal" times can be controlled in the human-involved workplace.
- Identify opportunities to rebalance the workload.
- Eliminate unlean practices wherever they are identified.
- Reduce worker idle times. Workers should not be waiting for work to do.
- Coordinate work input versus expected work output.
- Review and recalculate takt time regularly, particularly when work situations change (e.g., new operators, new equipment, new process, new design, new materials, etc.).

- Recognize that, sometimes, slowing down may lead to more effective utilization of human resources and a balanced efficiency of capital assets.
- If used properly, takt time helps to achieve *rhythm* in the workplace.

Resource work rate

Work rate and work time are essential components of estimating the cost of specific tasks. Thus, knowing worker work rates can be helpful for designing or redesigning work. Given a certain amount of work that must be done at a given work rate, the required time can be computed. Once the required time is known, the cost of the task can be computed on the basis of a specified cost-per-unit time. Work rate analysis is important for resource substitution decisions. The analysis can help identify where and when the same amount of work can be done with the same level of quality and within a reasonable time span by a less expensive resource. The results of learning curve analysis can yield valuable information about the expected work rate. The general relationship between work, work rate, and time is given by

$$\text{Work done} = (\text{work rate})(\text{time})$$

This is expressed mathematically as

$$w = rt$$

where:

w = the amount of actual work done expressed in appropriate units (e.g., miles of road completed, lines of computer code typed, gallons of oil spill cleaned, units of widgets produced, surface area painted)

r = the rate at which the work is accomplished (i.e., work accomplished per unit time)

t = the total time required to perform the work, excluding any embedded idle times

It should be noted that work is defined as a physical measure of accomplishment with uniform density. That means, for example, that one line of computer code is as complex and desirable as any other line of computer code. Similarly, cleaning one gallon of oil spill is as good as cleaning any other gallon of oil spill within the same work environment. The production of one unit of a product is identical to the production of any other unit of the product. If uniform work density cannot be assumed for the particular work being analyzed, then the relationship as presented may lead to erroneous conclusions. Uniformity can be enhanced if the scope of the analysis is limited to a manageable size. The larger the scope of the analysis, the more the variability from one work unit to another, and the less uniform the overall work measurement will be. For example, in a project involving the construction of 50 miles of surface road, the work analysis may be done in increments of 10 miles at a time rather than the total 50 miles. If the total amount of work to be analyzed is defined as one whole unit, then the relationship can be developed for the case of a single resource performing the work with the following parameters:

Resource:	Machine A
Work rate:	r
Time:	t
Work done:	100% (1.0)

The work rate, r, is the amount of work accomplished per unit time. For a single resource to perform the whole unit (100%) of the work, we must have the following:

$$rt = 1.0$$

For example, if machine A is to complete one work unit in 30 minutes, it must work at the rate of 1/30 of the work content per unit time. If the work rate is too low, then only a fraction of the required work will be performed. The information about the proportion of work completed may be useful for productivity measurement purposes. In the case of multiple resources performing the work simultaneously, the work relationship is as presented in Table 3.1.

Even though the multiple resources may work at different rates, the sum of the work they all perform must equal the required whole unit. In general, for multiple resources, we have the following relationship:

$$\sum_{i=1}^{n} r_i t_i = 1.0$$

Table 3.1 Work rate tabulation for multiple resources

Resource (i)	Work rate (r_i)	Time (t_i)	Work done (w)
RESource 1	r_1	t_1	$(r_1)(t_1)$
RESource 2	r_2	t_2	$(r_2)(t_2)$
…	…	…	…
RESource n	r_n	t_n	$(r_n)(t_n)$
		Total	1.0

where:

n = the number of different resource types
r_i = the work rate of resource type i
t_i = the work time of resource type i

For partial completion of work, the relationship is

$$\sum_{i=1}^{n} r_i t_i = p$$

where p is the proportion of the required work actually completed.

Work rate examples

Machine A, working alone, can complete a given job in 50 minutes. After machine A has been working on the job for 10 minutes, machine B was brought in to work with machine A in completing the job. Both machines working together finished the remaining work in 15 minutes. What is the work rate for machine B?

Solution

- The amount of work to be done is 1.0 whole units.
- The work rate of machine A is 1/50.
- The amount of work completed by machine A in the 10 minutes it worked alone is (1/50)(10) = 1/5 of the required total work.
- Therefore, the remaining amount of work to be done is 4/5 of the required total work.

Table 3.2 shows the two machines working together for 15 minutes. The computation yields:

$$\frac{15}{50} + 15(r_2) = \frac{4}{5}$$

which yields $r_2 = 1/30$. Thus, the work rate for machine B is 1/30. That means machine B, working alone, could perform the same job in 30 minutes.

Table 3.2 Work rate tabulation for machines A and B

Resource (i)	Work rate (r_i)	Time (t_i)	Work done (w)
Machine A	1/50	15	15/50
Machine B	r_2	15	$15(r_2)$
		Total	4/5

Table 3.3 Incorporation of resource cost into work rate analysis

Resource (i)	Work rate (r_i)	Time (t_i)	Work done (w)	Pay rate (p_i)	Pay (P_i)
Machine A	r_1	t_1	$(r_1)(t_1)$	p_1	P_1
Machine B	r_2	t_2	$(r_2)(t_2)$	p_2	P_2
…	…	…	…	…	…
Machine n	r_n	t_n	$(r_n)(t_n)$	p_n	P_n
		Total	1.0		Budget

In this example, it is assumed that both machines produce an identical quality of work. If quality levels are not identical, then the project analyst must consider the potentials for quality/time trade-offs in performing the required work. The relative costs of the different resource types needed to perform the required work may be incorporated into the analysis, as shown in Table 3.3.

Using the preceding relationship for work rate and cost, the work crew can be analyzed to determine the best strategy for accomplishing the required work, within the required time, and within a specified budget.

For another simple example of possible application scenarios, consider a case where an IT technician can install new IT software on three computers every 4 hours. At this rate, it is desired to compute how long it would take the technician to install the same software on five computers. We know, from the information given, that the proportion of three computers to 4 hours is the proportion of five computers to x hours, where x represents the number of hours the technician would take to install software on five computers. This gives the following ratio relationship:

$$\frac{3 \text{ computers}}{4 \text{ hours}} = \frac{5 \text{ computers}}{x \text{ hours}}$$

which simplifies to yield $x = 6$ hours and 40 minutes. Now consider a situation where the technician's competence with the software installation degrades over time for whatever reason. We will see that the time requirements for the IT software installation will vary depending on the current competency level of the technician. Learning curve analysis can help to capture such situations so that an accurate work time estimate can be developed.

Multiple work rate analysis

When multiple resources work concurrently at different work rates, the amount of work accomplished by each may be computed by the procedure for work rate analysis.

Example

Suppose the work rate of RES 1 is such that it can perform a certain task in 30 days. It is desired to add RES 2 to the task so that the completion time of the task can be reduced. The work rate of RES 2 is such that it can perform the same task alone in 22 days. If RES 1 has already worked 12 days on the task before RES 2 comes in, find the completion time of the task. Assume that RES 1 starts the task at time 0.

Solution

- The amount of work to be done is 1.0 whole units (i.e., the full task).
- The work rate of RES 1 is 1/30 of the task per unit time.
- The work rate of RES 2 is 1/22 of the task per unit time.
- The amount of work completed by RES 1 in the 12 days it worked alone is (1/30)(12) = 2/5 (or 40%) of the required work.
- Therefore, the remaining work to be done is 3/5 (or 60%) of the full task.
- Let T be the time for which both resources work together.

The two resources working together to complete the task yield Table 3.4.Thus, we have:

$$T/30 + T/22 = 3/5$$

which yields $T = 7.62$ days. Thus, the completion time of the task is $(12 + T) = 19.62$ days from time 0. The results of this example are summarized graphically in Figure 3.2. It is assumed that both resources produce an identical quality of work and that the respective work rates remain consistent. The respective costs of the different resource types may be incorporated into the work rate analysis. The critical resource diagramming (CRD) and resource schedule (RS) charts are simple extensions of very familiar tools. They are simple to use and they convey resource information quickly. They can be used to complement existing resource management tools. Users can find innovative ways to modify or implement them for specific resource planning, scheduling, and control purposes. For example, resource-dependent task durations

Table 3.4 Tabulation of resource work rates for RES 1 and RES 2

Resource type (i)	Work rate (r_i)	Time (t_i)	Work done (w_i)
RES 1	1/30	T	$T/30$
RES 2	1/22	T	$T/22$
		Total	3/5

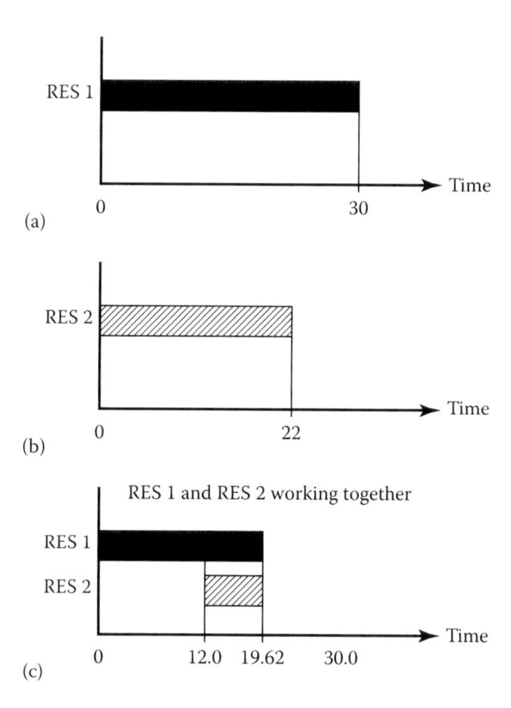

Figure 3.2 Resource schedule charts for RES 1 and RES 2. (a) RES 1 working alone. (b) RES 2 working alone. (c) RES 1 and RES 2 working together.

and resource cost can be incorporated into the CRD and RS procedures to enhance their utility for resource management decisions.

Resource loading graph

A worker is a crucial resource for an organization. Proper work design to better utilize workers is a key aspect of organization resource management. In this section, a worker is modeled as a resource. Resource loading refers to the allocation of workers to work elements. A resource loading graph is a graphical representation of resource allocation over time. A resource loading graph may be drawn for the different resource types involved in a project. Such a graph would provide information useful for work planning and budgeting purposes. A resource loading graph gives an indication of the demand a project will place on an organization's resources. In addition to resource units committed to activities, the graph may also be drawn for the other tangible and intangible resources

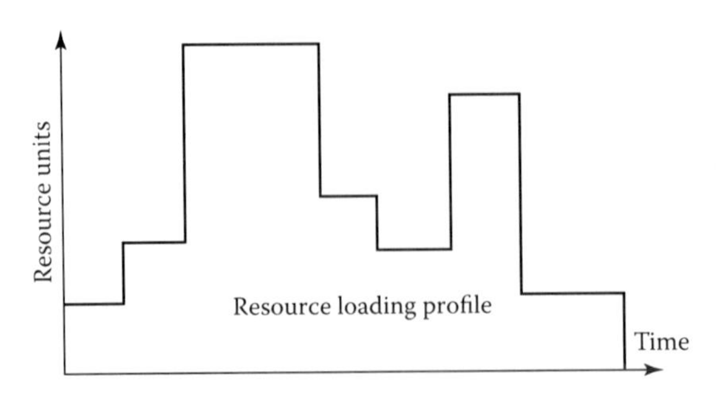

Figure 3.3 Generic profile of a resource loading graph.

of an organization. For example, a variation of the graph may be used to present information about the depletion rate of the budget available for a project. If drawn for multiple resources, it can help identify potential areas of work conflict. For situations where a single worker is assigned to multiple tasks, a variation of the resource loading graph can be developed to show the level of load (responsibilities) assigned to the worker over time. Figure 3.3 illustrates what a resource loading graph might look like.

Resource leveling

Resource leveling refers to the process of reducing the period-to-period fluctuation in a resource loading graph. If resource fluctuations are beyond acceptable limits, actions are taken to move activities or resources around in order to level out the resource loading graph. For example, it is bad for employee morale and public relations when a company has to hire and lay people off indiscriminately. Proper resource planning will facilitate a reasonably stable level of workforce. Other advantages of resource leveling include simplified resource tracking and control, lower resource management costs, and improved opportunities for learning. Acceptable resource leveling is typically achieved at the expense of longer project duration or higher project cost. Figure 3.4 shows a leveled resource loading.

When attempting to level resources, note that

1. Not all of the resource fluctuations can be eliminated.
2. Resource leveling often leads to an increase in project duration.

Resource leveling attempts to minimize fluctuations in resource loading by shifting activities within their available slacks. For small networks,

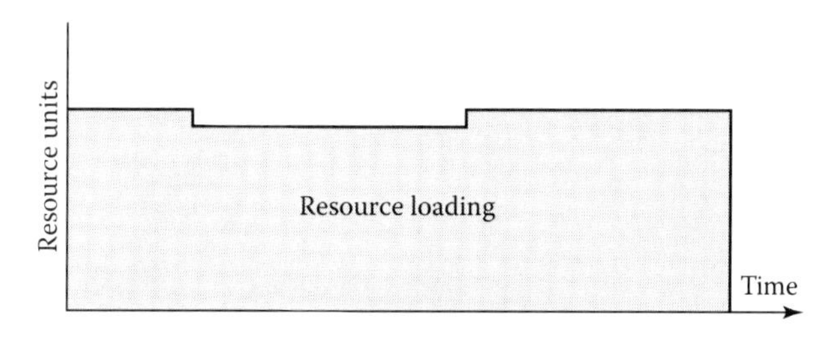

Figure 3.4 Balanced or leveled resource loading graph.

resource leveling can be attempted manually through trial-and-error procedures. For large networks, resource leveling is best handled by computer software techniques. Most of the available commercial project management software packages have internal resource leveling routines. One heuristic procedure for leveling resources, known as *Burgess's method*, is based on the technique of minimizing the sum of squares for the resource requirements in each period.

Assessment of work gaps

An assessment of work gaps can be done using a resource idleness graph, which is similar to a resource loading graph except that it is drawn for the number of unallocated resource units (or the number of idle workers) over time. The area covered by the resource idleness graph may be used as a measure of the effectiveness of the work-scheduling strategy of an organization. Suppose two scheduling strategies yield the same project duration, and suppose a measure of the resource utilization under each strategy is desired as a means to compare the strategies. Using a graph similar to that shown in Figure 3.5, we can assess the alternate strategies. The resource-idleness periods are computed as geometric areas as follows:

$$\text{Area A} = 6(5)+10(5)+7(8)+15(6)+5(16) = 306 \text{ resource} - \text{units} - \text{time}$$

$$\text{Area B} = 5(6)+10(9)+3(5)+6(5)+3(3)+12(12) = 318 \text{ resource} - \text{units} - \text{time}$$

Since area A is smaller than area B, it is concluded that strategy A is more effective for resource utilization than strategy B. Similar measures can be developed for multiple resources. However, for multiple resources, the different resource units must all be scaled to dimensionless quantities before computing the areas bounded by the resource idleness graphs.

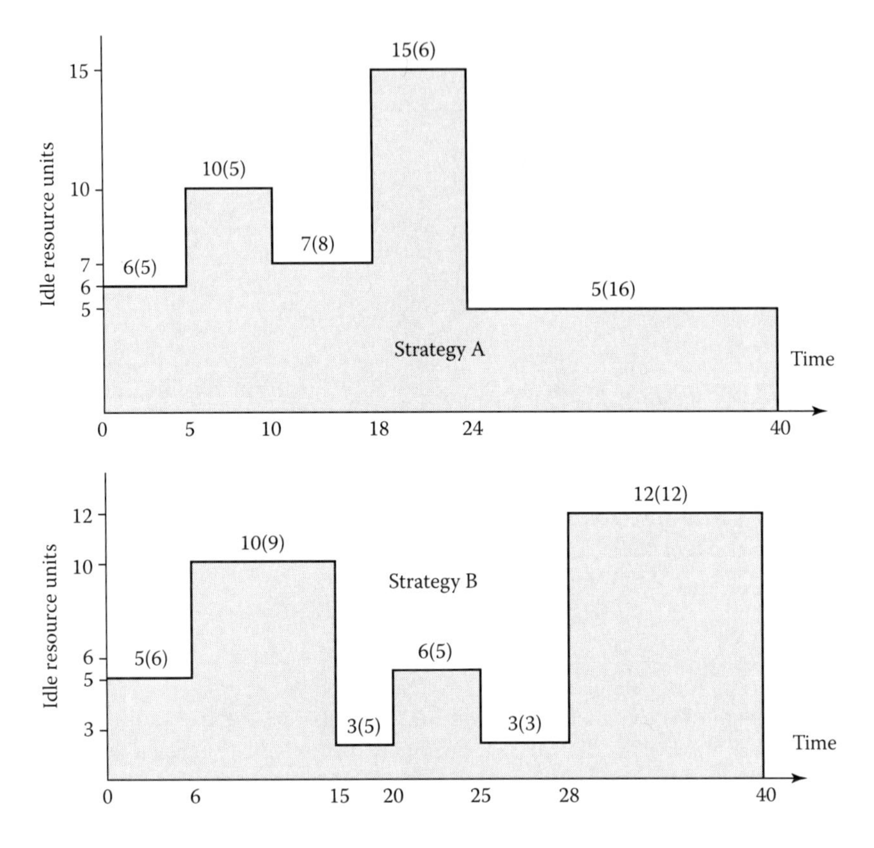

Figure 3.5 Resource idleness graphs for resource allocation.

Summary

In this chapter, we have presented a variety of topics related to using learning curves and techniques associated with design work. The conventional view of learning curves considers only one factor at a time as the major influence on productivity improvement. But in today's complex work environment, it is obvious that several factors interact to activate and perpetrate productivity improvement. Thus, multifactor learning curve models are useful for various types of analyses in the work environment. Such analyses include manufacturing economic analysis, break-even analysis, make-or-buy decisions, manpower scheduling, production planning, budgeting, resource allocation, labor estimating, and cost optimization. Multifactor learning curves can generate estimates about expected cost, productivity, process capability, work load composition, system response time, and so on. Such estimates can be valuable to decision makers for work design, process improvement, and work simplification. The availability of reliable learning curve estimates can enhance the communication interface between different groups in an organization. Multifactor learning curves can provide guidance for the effective cost-based implementation of design systems, facilitate the systems integration of production plans, improve supervisory interfaces, enhance design processes, and provide cost information to strengthen the engineering and finance interface. Work rate analysis, centered on work measurement, helps identify areas where operational efficiency and improvement can be pursued. In many production settings, workers encounter production breaks that require an analysis of the impact on production output. Whether production breaks are standard and scheduled or nonstandard and unscheduled, the workstation is subject to work rate slowdown (ramp-down) and work rate pickup (ramp-up), respectively, before and after a break. These impacts are subtle and are hardly noticed unless a formal, engineered work measurement is put in place.

References

Aft, L., Don't abandon your work measurement cap, *Industrial Engineering*, March 2010, Vol. 42, No. 3, pp. 37–40.

Asher, H., Cost-quantity relationships in the airframe industry, report no. R-291, Rand Corporation, Santa Monica, CA, July 1, 1956.

Badiru, A. B., Long live work measurement, *Industrial Engineer*, March 2008, Vol. 40, No. 3, p. 24.

Badiru, A. B., *Project Management in Manufacturing and High Technology Operations*, Wiley, New York, 1988.

Baloff, N., Extension of the learning curve: Some empirical results, *Operations Research Quarterly*, Vol. 22, No. 4, 1971, pp. 329–40.

Belkaoui, A., *The Learning Curve*, Quorun Books, Westport, CT, 1986.

Carlson, J. G. H., Cubic learning curves: Precision tool for labor estimating, *Manufacturing Engineering and Management*, Vol. 71, No. 5, 1973, pp. 22–25.

Carr, G. W., Peacetime cost estimating requires new learning curves, *Aviation*, Vol. 45, April 1946, pp. 9–13.

Chase, R. B. and N. J. Aquilano, *Production and Operations Management*, Richard E. Irwin, Homewood, IL, 1981.

Conley, P., Experience curves as a planning tool, *IEEE Spectrum*, Vol. 7, No. 6, 1970, pp. 63–68.

Conway, R. W. and A. Schultz, Jr, The manufacturing progress function, *Journal of Industrial Engineering*, Vol. 1, 1959, pp. 39–53.

Dada, M. and K. N. Srikanth, Monopolistic pricing and the learning curve: An algorithmic approach, *Operations Research*, Vol. 38, No. 4, 1990, pp. 656–66.

DeJong, J. R., The effects of increasing skill on cycle time and its consequences for time standards, *Ergonomics*, Vol. 1, November 1957, pp. 51–60.

Ernst, C. and A. Szczesny, Cost accounting implications of surgical learning in the DRG era: Data evidence from a German hospital, *Schmalenbach Business Review*, Vol. 57, No. 3, April 2005, pp. 127–66.

Glover, J. H., Manufacturing progress functions: An alternative model and its comparison with existing functions, *International Journal of Production Research*, Vol. 4, No. 4, 1966, pp. 279–300.

Grahame, S., The learning curve. The key to future management?, *Research Executive Summary Series*, Vol. 6, No. 12, 2010, pp. 1–10.

Hirchmann, W. B., Learning curve, *Chemical Engineering*, Vol. 71, No. 7, 1964, pp. 95–100.

Jaber, M. Y. and A. L. Guiffrida, Learning curves for imperfect production processes with reworks and process restoration interruptions, *European Journal of Operational Research*, Vol. 189, No. 1, 2008, pp. 93–104.

Jewell, W. S., A generalized framework for learning curve reliability growth models, *Operations Research*, Vol. 32, No. 3, May–June 1984, pp. 547–58.

Knecht, G. R., Costing, technological growth and generalized learning curves, *Operations Research Quarterly*, Vol. 25, No. 3, September 1974, pp. 487–91.

Kopcso, D. P., Learning curves and lot sizing for independent and dependent demand, *Journal of Operations Management*, Vol. 4, No. 1, November 1983, pp. 73–83.

Levy, F. K., Adaptation in the production process, *Management Science*, Vol. 11, No. 6, April 1965, pp. B136–54.

Liao, W. M., Effects of learning on resource allocation decisions, *Decision Sciences*, Vol. 10, 1979, pp. 116–25.

Pegels, C. C., On startup or learning curves: An expanded view, *AIIE Transactions*, Vol. 1, No. 3, September 1969, pp. 216–22.

Richardson, W. J., Use of learning curves to set goals and monitor progress in cost reduction programs, *Proceedings of 1978 IIE Spring Conference*, Institute of Industrial Engineers, Norcross, GA, 1978, pp. 235–39.

Smith, J., *Learning Curve for Cost Control*, Industrial Engineering & Management Press, Norcross, GA, 1989.

Smunt, T. L., A comparison of learning curve analysis and moving average ratio analysis for detailed operational planning, *Decision Sciences*, Vol. 17, No. 4, 1986, pp. 475–95.

Steven, G. J., The learning curve: From aircraft to spacecraft, *Management Accounting*, London, Vol. 77, No. 5, 1999, pp. 64–73.

Teplitz, C. J. and J.-P. Amor, An efficient approximation for project composite learning curves, *Project Management Journal*, Vol. 29, No. 3, 1998, pp. 34–42.

Towill, D. R. and U. Kaloo, Productivity drift in extended learning curves, *Omega*, Vol. 6, No. 4, 1978, pp. 295–304.

Waller, E. W. and T. J. Dwyer, Alternative techniques for use in parametric cost analysis, *Concepts: Journal of Defense Systems Acquisition Management*, Vol. 4, No. 2, Spring 1981, pp. 48–59.

Wilson, J. R. and N. Corlett, eds, *Evaluation of Human Work*, 3rd Edition, Taylor and Francis, New York, 2005.

Wright, T. P., Factors affecting the cost of airplanes, *Journal of Aeronautical Science*, Vol. 3, No. 2, February 1936, pp. 122–28.

Yelle, L. E. Adding life cycles to learning curves, *Long Range Planning*, Vol. 16, No. 6, December 1983, pp. 82–87.

Yelle, L. E., Estimating learning curves for potential products, *Industrial Marketing Management*, Vol. 5, No. 23, June 1976, pp. 147–54.

Yelle, L. E., The learning curve: Historical review and comprehensive survey, *Decision Sciences*, Vol. 10, No. 2, April 1979, pp. 302–28.

section three

Work evaluation

chapter four

Work performance measures

> Structured methods for planning, measuring, and executing work greatly increase the likelihood of positive outcomes.
>
> **Ibrahim Badiru,**
> *General Motors, October 29, 2016 (Badiru, 2016)*

Work is defined simply as actions to accomplish an end goal. Such actions may be temporary, intermittent, or recurring. In each case, a performance assessment is essential for ensuring that the actions follow a path to the desired end goal. Qualitative and quantitative performance measures are necessary to guide decision makers, managers, supervisors, and the workers themselves. While tools and techniques for work performance measures have been available in the literature for a long time (Badiru, 1994; DOE, 1995; Sieger and Badiru, 1995), there is limited guidance on how and when to apply them consistently. In this book, modernized tools, techniques, and guides are provided to facilitate expedient and consistent applications of generalized performance measures in any work environment. There are internal customers (the organization) and external customers (outside clients) of work performance outputs. The needs and expectations of all parties must be taken into consideration when embarking on specific performance measure techniques. Employee performance measures have direct impacts on quality of work and organizational process improvement efforts, as has been documented in the literature over the years (Harrington, 1991; Juran, 1964, 1988, 1989; Juran and Gryna, 1980, Juran Institute, 1989, 1990; Pall, 1987; Badiru and Ayeni, 1993).

Performance measures are recognized as an important element of all work improvement programs. Managers and supervisors directing the efforts of an organization or a group have a responsibility to know how, when, and where to institute a wide range of changes. These changes cannot be sensibly implemented without knowledge of the appropriate information on which they are based. Many organizations have no standardized approach to developing and implementing performance measurement systems. As a result, performance measures are not often fully adopted to gauge the success of the various organizational improvement programs in both small and large organizations. Measuring performance can be complex, time consuming, controversial, contentious, and inconsistent. Thus, some managers, leaders, and supervisors default to the *do-nothing* mode of performance assessment. Based on the available literature (e.g., DOE, 1995; Geisler, 2012), this chapter provides an updated, comprehensive, and step-by-step process of how to develop work performance measurements at any level within an organization and how to evaluate the effectiveness of the performance measures. One key purpose of performance measurement is to give feedback in order to improve work. The best feedback should meet the following requirements:

- Be concise.
- Be intentional.
- Be ongoing.
- Be specific.
- Be focused on the performance outcome.
- Be tailored to the individual and work element involved.
- Be interactive (listening and talking).
- Be timely.

The implementation of performance measurements for a specific process should involve as many employees and stakeholders as possible to stimulate ideas and reinforce the notion that this is a team effort requiring buy-ins from all involved in order to succeed. Substantial benefits are realized by organizations implementing performance measurement programs at a grassroots level. The benefits are realized by organizations implementing performance measurement programs almost immediately through an improved understanding of processes by all employees, from top to bottom. Furthermore, individuals get an opportunity to receive a broadened perspective of the organization's functions, rather than the more limited perspective of their own immediate work environment and span of control.

As a process, performance measurement is not simply concerned with collecting data associated with a predefined performance goal or standard. Performance measurement is better thought of as an overall management system involving prevention and detection aimed at achieving conformance of the work product or service to the organization's requirements. Additionally, it is concerned with process optimization through the increased efficiency and effectiveness of the process or product. These actions occur in a continuous cycle, allowing options for the expansion and improvement of the work process or product as better techniques are discovered and implemented, particularly where learning and relearning are involved (Badiru and Ijaduola, 2009). Performance measurement is primarily managing outcome, and one of its main purposes is to reduce or eliminate overall variation in the work product or process. The goal is to arrive at sound decisions about actions affecting the product or process and its output.

Reemergence of work measurement

Innovation and advancement owe their sustainable foundations to some measurement scale. We must measure work before we can improve it. This means we *still* need work measurement. Work measurement is precisely what is needed as an input to enhance work design.

One common thread in organizational pursuits is the goal of achieving improvements in something, be it a product, a process, a service, or results. How can we improve anything if we don't know how long it takes us to accomplish the task in the first place? Industrial work measurement has reemerged in new improved forms, names, and analytics. Six Sigma, lean, total quality management (TQM), 5S, the theory of constraints, supply networking, and so on all have the same underlying principles of improving work elements. Even in personal pursuits, time and motion analyses, though subconsciously, do play a role in charting the path to effective actions. The most successful leaders are those whose visions contain elements of time, motion toward action, and deliberate work standards for their employees.

A common adage says, "You cannot control what you cannot measure." If you cannot control it, it means that you cannot improve it. This means that work measurement is essential. The fast-changing operating environments that we face nowadays necessitate interdisciplinary improvement projects that may involve physicists, material scientists, chemists, engineers, bioscientists, and a host of other disciplines. They may all use different measurement scales and communicate in different jargons. But the common thread of work measurement permeates their collective improvement efforts. If we review the widely accepted definitions of engineering, we will see references to the need to measure work—as in energy measurement, cost measures, productivity measurement, scientific measurements, sensory data, and systems capability assessment. These present convincing confirmation that work measurement has reemerged in more contemporary concatenations of words.

Measurement of productivity, human performance, and resource consumption are essential components of achieving organizational goals and increasing profitability. For example, trade-offs must be exercised among the triple constraints of production represented in terms of time, cost, and performance. Proper work measurement is required to ensure that the trade-offs are balanced and done within feasible regions of the operating environment. Work rate analysis, centered on work measurement, helps to identify areas where operational efficiency and improvement can be pursued. In many production settings, workers encounter production breaks that require an analysis of the impact on production output. Some breaks are standard and scheduled breaks, while others are nonstandard and unscheduled occurrences. In each case, the workstation is subject to work rate slowdown (ramp-down) and work rate pickup (ramp-up), respectively before and after a break. These impacts are subtle and are hardly noticed unless a formal, engineered work measurement is put in place with appropriate performance measures.

Determination of performance measures

The appropriate performance measures must be determined for the appropriate work. Performance measures quantitatively reveal something important about products, services, and the processes that produce them. Performance measures facilitate understanding, managing, and improving organizational work efforts. Performance measures reveal the following:

- How well a function is doing
- If goals are being met
- If constituents and stakeholders are satisfied
- If work processes are in statistical control
- If and where improvements are necessary and possible

Performance measures provide the information necessary to make intelligent decisions about work functions and responsibilities. A performance measure is composed of a number and a unit of measure. The number gives us a magnitude (how much) and the unit gives the number a relevant meaning (what) in the context of the work scenario. Performance measures are always tied to a goal or an objective (the target). Performance measures can be represented by single-dimensional units such as hours, meters, nanoseconds, dollars, number of reports, number of errors, number of professionally certified employees, length of time to design hardware, and so on. They can show the variation in a process or deviation from design specifications. Single-dimensional units of measure usually represent very basic and fundamental measures of some process or product.

Every performance decision requires data collection, measurement, and analysis. In practice, there are different types of measurement scales depending on the particular items of interest. Data may need to be collected on decision factors, costs, performance levels, outputs, and so on. The different types of data measurement scales that are applicable for performance assessment include nominal scales, ordinal scales, interval scales, and ratio scales.

A *nominal scale* is the lowest level of measurement scales. It classifies items into categories. The categories are mutually exclusive and collectively exhaustive. That is, the categories do not overlap and they cover all possible categories of the characteristics being observed. For example, in the analysis of the critical path in a project network, each job is classified as either critical or noncritical. Gender, type of industry, job classification, and color are examples of measurements on a nominal scale.

An *ordinal scale* is distinguished from a nominal scale by the property of order among the categories. An example is the process of prioritizing tasks for resource allocation. We know that first is above second, but we do not know how far above. Similarly, we know that better is preferred to good, but we do not know by how much. In quality control, the ABC classification of items based on the Pareto distribution is an example of a measurement on an ordinal scale.

An *interval scale* is distinguished from an ordinal scale by having equal intervals between the units of measurement. The assignment of priority ratings to project objectives on a scale of 0–10 is an example of a measurement on an interval scale. Even though an objective may have a priority rating of zero, it does not mean that the objective has absolutely no significance to the project team. Similarly, the scoring of zero on an examination does not imply that a student knows absolutely nothing about the materials covered by the examination. Temperature is a good example of an item that is measured on an interval scale. Even though there is a zero point on the temperature scale, it is an arbitrary relative measure. Other examples of interval scales are IQ measurements and aptitude ratings.

A *ratio scale* has the same properties as an interval scale but with a true zero point. For example, an estimate of zero time units for the duration of a task is a ratio scale measurement. Other examples of items measured on a ratio scale are cost, time, volume, length, height, weight, and inventory level. Many of the items measured in engineering systems will be on a ratio scale.

An important aspect of performance measurement involves the classification scheme used. Most systems will have both quantitative and qualitative data. Quantitative data require that we describe the characteristics of the items being studied numerically. Qualitative data, on the other hand, are associated with attributes that are not measured numerically. Most items measured on the nominal and ordinal scales will normally be classified into the qualitative data category, while those measured on the interval and ratio scales will normally be classified into the quantitative data category. The implication for performance control is that qualitative data can lead to bias in the control mechanism because qualitative data are subject to the personal views and interpretations of the person using the data. As much as possible, data for performance measurement and management should be based on a quantitative measurement.

More often, multidimensional units of measure are used. These are performance measures expressed as ratios of two or more fundamental units. These may be units such as miles per gallon (a performance measure of fuel economy), the number of accidents per million hours worked (a performance measure of the companies safety program), or the number of on-time vendor deliveries per total number of vendor deliveries. Performance measures expressed this way almost always convey more information than the single-dimensional or single-unit performance measures. Ideally, performance measures should be expressed in units of measure that are the most meaningful to those who must use or make decisions based on those measures.

Most performance measures can be grouped into one of the following six general categories (DOE, 1995). However, certain organizations may develop their own categories as appropriate depending on the organization's mission.

1. *Effectiveness*: A process characteristic indicating the degree to which the process output (work product) conforms to requirements. That is, are we doing the right things?
2. *Efficiency*: A process characteristic indicating the degree to which the process produces the required output at minimum resource cost. That is, are we doing the right things right?
3. *Quality*: The degree to which a product or service meets customer requirements and expectations.
4. *Timeliness*: Measures whether a unit of work was done correctly and on time. Criteria must be established to define what constitutes timeliness for a given unit or work. The criterion is usually based on customer requirements.
5. *Productivity*: The value added by the process divided by the value of the labor and capital consumed.
6. *Safety*: Measures the overall health of the organization and the working environment of its employees.

The following reflect the attributes of an ideal unit of measure.

- Reflects the user's needs as well as the organization's needs
- Provides an agreed basis for decision making
- Is understandable
- Applies broadly
- May be interpreted uniformly
- Is compatible with existing methods (a way to measure it exists)
- Is precise in interpreting the results
- Is economical to apply

Performance data must support the mission assignment(s) from the highest organizational level down to the performance level. Therefore, the measurements that are used must reflect the assigned work at the level. Within a system, units of measure should interconnect to form a pyramid, whereby the technological units start at the base. These are measures of individual units of products and of individual elements of service. The next level of units serves to summarize the basic data (e.g., percent defective for specific processes, documents, product components, service cycles, and persons).

Next are units of measure that serve to express quality for entire departments, product lines, and classes of service. In large organizations, there may be multiple layers of this category.

At the top are the financial and upper management units (measures, indexes, ratios, etc.), which serve the needs of the highest levels in the organization: corporate, divisional, and functional. Elements of the pyramid include the following items, which may be applied to the pyramid based on each organization's internal practices and managerial preferences.

- Corporate reports: money, ratios, index
- Measures of broad matters such as quality compared with that of competitors; time required to launch new products
- Measures that help to establish departmental quality goals and to evaluate departmental performance against goals
- Technological units of measure for individual elements of the product, process, or service
- Managerial indicators consisting mostly of composites of data expressed in such forms as summaries, ratios, and indices
- Upper management indicators to evaluate broad matters include data systems, reports, audits, and personal observations
- Department summaries of product and process performance, derived from inspections and tests, reports of nonconformance, and so on, and personal observations
- Numerous technological instruments to measure technological product and process features

Benefits of measurements

The following are seven important benefits of measurements.

1. To identify whether customer requirements are being met—that is, whether an organization is providing the services and products that customers require.
2. To help an organization understand its processes; to confirm what the organization knows or to reveal what is not known. Does the organization know where the problems originated or are located?
3. To ensure decisions are based on fact, not on emotion. Are decisions based on well-documented facts and figures or on intuition and gut feelings?
4. To show where improvements need to be made. Where can the organization do better? How can improvements be pursued?

5. To show if improvements actually happened. Does the organization have a clear picture of the improvement and resulting impacts?
6. To reveal problems related to bias, emotion, and longevity in the organization. If an organization has been doing its job for a long time without measurements, it might be assumed incorrectly that things are going well.
7. To identify whether suppliers are meeting requirements. Do suppliers know if the organization's requirements are being met?

Justification for work performance measurement

If an activity cannot be measured, it cannot be controlled. If an activity cannot be controlled, it cannot be managed. Without dependable measurements, intelligent decisions cannot be made. Measurements, therefore, can be used for the following purposes:

1. *Control*: Measurements help to reduce variation. For example, a typical control for contractor accountability measurement might be a work authorization document evaluated against a performance evaluation plan, which has the purpose of reducing expense overruns so that agreed objectives can be achieved.
2. *Self-assessment*: Measurements can be used to assess how well a process is doing, including improvements that have been made.
3. *Continuous improvement*: Measurements can be used to identify defect sources, process trends, and prevent defects, and to determine process efficiency and effectiveness, as well as opportunities for improvement.
4. *Management assessment*: Without measurements, there is no way to be certain if value-added objectives are being met or that a process is effective and efficient. The basic concept of performance measurement involves (a) planning and meeting established operating goals/standards, (b) detecting deviations from planned levels of performance, and (c) restoring performance to the planned levels or achieving new levels of performance.

Foundation for good performance measurement

Successful performance measurement systems adhere to the following principles:

1. Measure only what is important. Do not measure too much; measure things that impact the work outcome.
2. Focus on organizational needs. Ask your work constituents if they think this is what should be measured.

3. Involve workers in the design and implementation of the measurement system. Give workers a sense of ownership and empowerment, which lead to improvements in the quality of the measurement system.

The basic feedback loop shown in Figure 4.1 presents a systematic series of steps for maintaining conformance to goal/standards by communicating performance data back to the responsible worker and/or decision maker to take appropriate actions(s).

Without the basic feedback loop, no performance measurement system will ever ensure an effective and efficient operation, and, as a result, conformance to customers' requirements.

The message of the feedback loop is that to achieve the goal or standard, those responsible for managing the critical activities must always be in a position to know (1) what is to be done, (2) what is being done, (3) when to take corrective actions, and (4) when to change the goal or standard. The basic elements of the feedback loop and their interrelations are as follows:

1. The *indicator* evaluates actual performance.
2. The indicator reports this performance to a *responsible worker*.
3. The responsible worker also receives information on what the goal or standard is.
4. The responsible worker compares actual performance with the goal. If the difference warrants action, the worker reports to a *responsible decision maker*. (This could signal a need for correction action.)

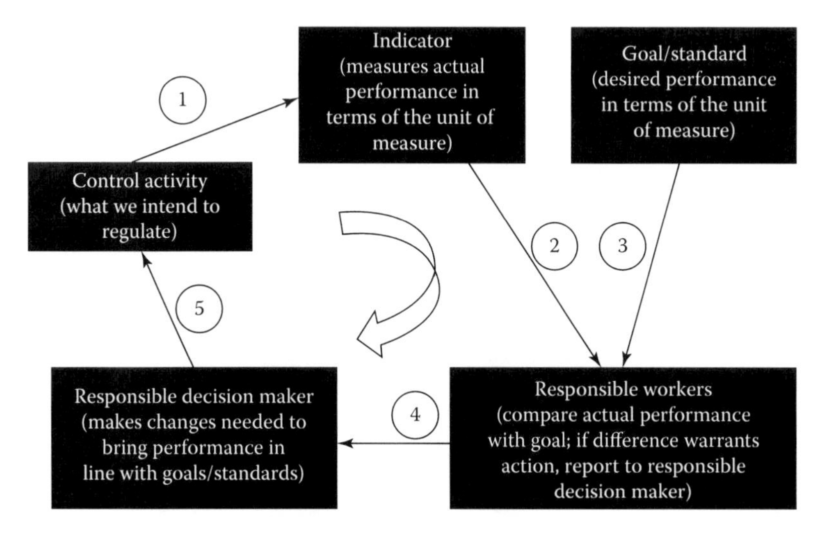

Figure 4.1 Basic work feedback loop.

5. The responsible decision maker verifies variance, determines if corrective action is necessary, and, if so, makes the changes needed to bring performance back in line with the goals.

Process overview

Figure 4.2 shows a high-level block diagram of the performance measurement process. It has been separated into 11 discrete steps. This is a guideline, intended to show the process generically. Different organizations, who best know their own internal processes, should feel free to adapt the guidelines where necessary to best fit within their operations. Subcomponents within the steps may need to be exchanged, or it may be necessary to revisit previously completed steps of the process based on new information arising from latter steps.
A brief description of each of the process steps follows:

1. *Identify the process flow*. This is the first and perhaps most important step. If your employees cannot agree on their process(es), how can they effectively measure them or utilize the output of what they have measured?

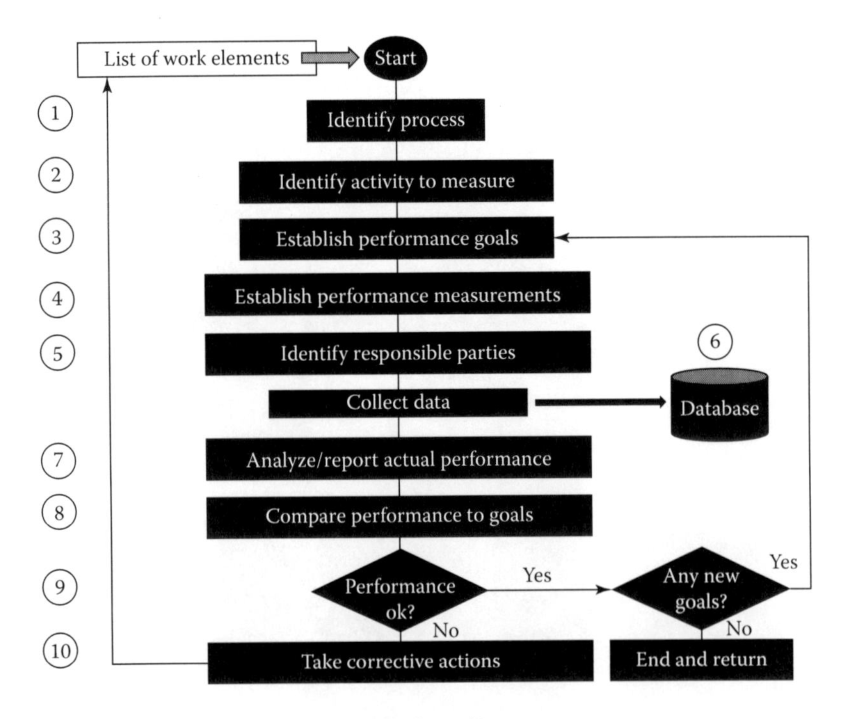

Figure 4.2 Performance measurement flow diagram.

2. *Identify the critical activity to be measured*. This is the culminating activity at which it makes the most sense to locate an indicator and define an individual performance measure within the process.
3. *Establish performance goals or standards*. All performance measures should be tied to a predefined goal or standard, even if the goal is at first somewhat subjective. Having goals and standards is the only way to meaningfully interpret the results of your measurements and gauge the success of your management systems.
4. *Establish performance measurements*. In this step, you continue to build the performance measurement system by identifying individual measures.
5. *Identify responsible parties*. A specific entity (as in a team or an individual) needs to be assigned the responsibilities for each of the steps in the performance measurement process.
6. *Collect data*. In addition to writing down the numbers, the data need to be preanalyzed in a timely fashion to observe any early trends and confirm the adequacy of your data collection system.
7. *Analyze/report actual performance*. In this step, the raw data are formally converted into performance measures, displayed in an understandable form, and disseminated in the form of a report.
8. *Compare actual performance to goals*. In this step, compare performance, as presented in the report, with predetermined goals or standards and determine the variation (if any).
9. *Are corrective actions necessary?* Depending on the magnitude of the variation between measurements and goals, some form of corrective action may be required.
10. *Make changes to bring performance back in line with goals*. This step only occurs if corrective action is necessary. The actual determination of the corrective action is part of the quality improvement process, not the performance measurement process. This step is primarily concerned with the improvement of your management system.
11. *Are new goals needed?* Even in successful systems, changes may need to be revised in order to establish ones that challenge an organization's resources, but do not overtax them. Goals and standards need periodic evaluation to keep up with the latest organizational processes.

Step 1: Identify the process

In identifying the process, an understanding of *what* you want to measure is of critical importance. Usually there are many processes and functions, each potentially needing performance measures. If there are multiple processes, consider the business impacts and select those processes that are most important to the customer (both internal and external) to satisfy their requirements and/or those processes with problem areas identified by management. These then become the key (or important) processes.

A process needs to be manageable in size. A lot of effort can be wasted if you do not start with a well-defined process. You should ask the following questions:

1. What product or service do we produce?
2. Who are our customers?
3. What comprises our process?
   a. What do we do?
   b. How do we do it?
   c. What starts our process?
   d. What ends our process?

Before you try to control a process, you must understand it. A flow diagram is an invaluable tool and the best way to understand a process. Flowcharting the entire process, down to the task level, sets the stage for developing performance measures.

All parties who are involved in the process should participate in creating the flowcharts. In a team environment, individuals will receive a new understanding of their processes. As participants, you can count on their later support to make the performance measurement system work.

Step 2: Identify critical activity/activities to be measured

It is important to choose only the critical activity/activities to be measured. We measure these activities to control them. Controlling, or keeping things on course, is not something we do in the abstract. Control is applied to a specific critical activity. When making your selection, focus on key areas and processes rather than people.

Examine each activity in the process and identify those that are critical. Critical activities are those that significantly impact total process efficiency, effectiveness, quality, timeliness, productivity, or safety. At the management level, critical activities impact management priorities, organizational goals, and external customer goals.

Ask the question, Does it relate, directly or indirectly, to the ultimate goal of customer satisfaction? Every critical activity should. For example, on-time delivery is directly related to customer satisfaction. Use quality tools such as the Pareto principle, brainstorming, or examining date to help prioritize the critical activities.

Confirm that the activity is critical. Do all concerned agree that this activity needs to be watched closely and acted on if its performance is less than desirable? Is it something that should be continuously improved? Does the benefit exceed the cost of taking the measurement? If the answer is "no" to any of these questions, you should reevaluate why you consider it critical. Each critical activity becomes the hub around which a feedback loop is constructed (Figure 4.1).

It is at this step that we begin to think about what we want to know or understand about the critical activity and/or process. Perhaps the most fundamental step in establishing any measurement system is answering the question, what do I want? The key issue then becomes, how do we generate useful information? Learning to ask the right questions is a key skill in effective data collection. Accurate, precise data collected through an elaborately designed statistical sampling plan are useless if they do not clearly address a question that someone cares about. It is crucial to be able to state precisely what it is you want to know about the activity you are going to measure. Without this knowledge, there is no basis for making measurements. To generate useful information, planning for good data collection proceeds along the following lines:

- What question do we need to answer?
- How will we recognize and communicate the answers to the question?
- What data analysis tools (Pareto diagrams, histograms, bar graphs, control charts, etc.) do we envision using? How will we communicate the results?
- What type of data do the data analysis tools require?
- Where in the process can we get these data?
- Who in the process can give us these data?
- How can we collect these data from people with minimum effort and chance of error?
- What additional information do we need to capture for future analysis, reference, and tractability?

Notice how this planning process (Figure 4.3) essentially works backward through the model for generating useful information. We start by defining the question. Then, rather than diving into the details of data collection, we consider how we might communicate the answer to the question and what types of analysis we will need to perform. This helps us define our data needs and clarifies what characteristics are most important in the data. With this understanding as a foundation, we can deal more coherently with the where, who, how, and what else issues of data collection.

Information generating begins and ends with questions. To generate information, we need to

- Formulate precisely the question we are trying to answer
- Collect the data and facts relating to that question

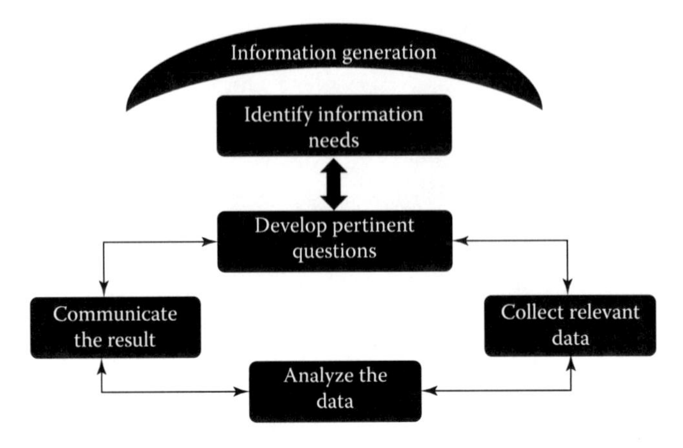

Figure 4.3 Model for generating useful information.

- Analyze the data to determine the factual answer to the question
- Present the data in a way that clearly communicates the answer to the question

Step 3: Establish performance goals or standards

Goals and standards are necessary; otherwise, there is no logical basis for choosing what to measure, what decisions to make, or what action to take. Goals can be a management directive or can be set in response to customer needs or complaints. Know your customers and their expectations. For each critical activity selected for measurement, it is necessary to establish a performance goal or standard. This is a target to be aimed at, an achievement toward which effort is expended. Standards often are mandated by external sources (e.g., the Occupational Safety and Health Administration [OSHA], government regulations, etc.). Knowledge of performance is not enough, you must have a basis for comparison before you can decide or act.

The concept of establishing performance goals/standards is not limited to numbered quantities (e.g., budget or deliveries). Neither is it limited to "things." The concept of standards extends to business practices, routines, methods, and procedures as well.

Performance goals can be established for (1) the overall process output and/or (2) the critical activities that produce the output. In any case, if this is the first set of goals or standards to be established and no basis for setting goals or standards exists, a baseline period of observation is appropriate prior to establishing the goal or standard.

Good performance goals or standards are

- *Attainable*: Should be met with reasonable effort under the conditions that are expected to prevail.

- *Economic*: The cost of setting and administering should be low in relation to the activity covered.
- *Applicable*: Should fit the conditions under which they are to be used. If conditions vary, they should contain built-in flexibility to meet these variables.
- *Consistent*: Should help to unify communication and operations throughout all functions of the company.
- *All-inclusive*: Should cover all interrelated activities. Failing this, standards will be met at the expense of those activities for which standards have not been set.
- *Understandable*: Should be expressed in simple, clear terms, so as to avoid misinterpretation or vagueness. Instructions for use should be specific and complete.
- *Measurable*: Should be able to communicate with precision.
- *Stable*: Should have a long enough life to provide predictability and to amortize the effort of preparing them.
- *Adaptable*: Should be designed so that elements can be added, changed, and brought up to date without redoing the entire structure.
- *Legitimate*: Should be officially approved.
- *Equitable*: Should be accepted as a fair basis for comparison by the people who have the job of meeting the goal or standard.
- *Customer focused*: Should address areas important to the customer (internal/external), such as cycle time, quality, cost schedule performance, and customer satisfaction.

Step 4: Establish performance measurements

This step involves performing several activities that will continue to build the performance measurement system. Each performance measurement consists of a defined unit of measure (the performance measure itself), an indicator to measure or record the raw data, and a frequency with which the measurements are made. To develop a measure, the team

- Translates the question "What do I want to know?" into a performance measure
- Identifies the raw data that will generate the performance measure
- Determines where to locate the raw data
- Identifies the indicator or measurement instrument that will collect the data for the performance measures
- Determines how often to take the measurements

At this point, your team has agreed on which process to measure (step 1); identified the critical activities of your process with emphasis on those that impact quality, efficiency, timeliness, customer satisfaction, and so on (step 2); looked at goals for these activities, products, and services (where they exist); and has quantified these goals where possible (step 3). Your team should use the knowledge gained from these previous steps to help state precisely what you want to know about the critical activities or the process as a whole. Think of this step as one that will allow you to generate useful information rather than just data. The purpose of this information is to provide everyone involved with an agreed basis for making sensible decisions about your processes, products, and services. Don't move on until the team agrees on what information you are trying to extract from the measurements.

Translate into performance measures

Having identified precisely what you want to know or understand about your process, you must now assemble this knowledge into a performance measure. Performance measures, and the data necessary to generate them, should be chosen to answer the questions you have just posed. At this point, your team must decide how you will "say it in numbers."

Performance measures are generally easiest to determine for activities or processes that have established and quantified goals. In such cases, the performance measures are usually stated in the same units as or similar units to the goals.

When no goals exist for an activity (or the process as a whole), the team should revisit the fundamental question of what it is they wish to know. The performance measures should provide quantitative answers to their questions in units that relate to those questions. The team may wish to reread the beginning of this chapter to reinforce the concept of a unit of measure and what it should convey.

The following example of a vendor selection process should prove useful in illustrating how to turn a posed question into a possible performance measure:

> You are part of a work team within the procurement department of your company. Over the years, the number of vendors from which you make purchases has grown astronomically and you need some basis on which to help decide which vendors perform the best. You have concluded that one of the more fundamental questions you would like to answer is, how well do our vendors meet the contract delivery dates? Your team needs to choose a performance measure that will help answer this question. After putting several possible performance measures on a flip chart and examining what information each

> could convey, the team decide to use the percentage of on-time deliveries per month.
>
> To ensure the team understand what the measure will provide them, they rewrite this measure in terms of the units that are actually used to calculate it. The performance measure then looks like this:
>
> $$\frac{\text{Number of on-time deliveries per month}}{\text{Total number of deliveries per month}} \times 100\%$$
>
> Both versions of the performance measure are essentially the same, but the second actually conveys more information to the reader and provides an indication of what data goes into the measurement. This performance measure should help the team answer the question of how well vendors are meeting contract delivery dates. By writing their performance measure in more fundamental units, the team will be better prepared to move to the next activity, which is identifying the raw data needed.

A good way to "test" a team's understanding of the performance measures they have chosen is to have them describe how they would display their results graphically. Have the team explain what type of graph they would use for each performance measure and how they would interpret the results. Quite often, seeing a performance measure displayed graphically will help determine if it will actually provide the information needed. Doing this simple step at this stage will ensure the team has chosen the right performance measure.

In reality, many work teams may find that some of their performance measures do not really tell them what they want to know. Don't panic, even performance measures that don't quite work may help refocus the team on the real issues they hope to address. Introduce a new set of measures and try again.

Identify raw data

The purpose of this activity is to identify the raw data you will need to generate the performance measures. It is difficult to perform a measurement if the required data and data source have not been identified. For simple processes with straightforward performance measures, this step may seem simple. However, very complex or high-level performance measures may require many raw data from numerous sources.

In general, performance measures are seldom generated directly in a single measurement or from a single source. They usually (but not always) consist of some combination of other raw data elements, as in the preceding example. To illustrate the difference, consider the following examples:

1. Your workgroup enters data from customer order forms into an electronic database. Your group decide that the number of errors per day is a useful performance measure for that process. The raw data for your measurement consist of the number of errors in the database each day. In this case, the collection of raw data *is* the performance measure and it has been measured directly.
2. You are in the procurement department and your team has decided to use the percentage of on-time deliveries per month of key vendors as a performance measure. The raw data you need consist of four sets. First, you need the delivery date on the contract that was awarded to your vendors. Second, you need the date the delivery was made. Third, you must compute whether the delivery was on time and count how many deliveries there were for each vendor. Fourth, the team will need the total number of deliveries made within the month for that vendor. Unlike example 1, several elements of raw data are required to reconstruct the performance measure.
3. Your management team considers the company's overhead burden rate to be an excellent high-level performance measure. This measure is very complex and is frequently performed by the company's accountants and budget analysts. Such measures require many raw data elements that consist of facilities costs, human resource benefits, training costs, rework costs, sales income, and so on. This performance measure requires that many lower-level measures are taken and *rolled up* into the higher-level measure. Many data elements must be collected along the way.

When the team completes this activity, it should have a list of the raw data elements needed to generate the performance measures. In addition, the team should consider what, if any, computations or calculations must be performed with the data.

Locate raw data

The purpose of this activity is to determine if and where the data exist. Stated differently, it is a matter of determining at which step in a process to take a measurement, at which point in time, or at which physical or geographical location. Quite often, this activity is performed concurrently with the previous one.

In the simplest case, you may find that your work group already has the raw data collected and you need only retrieve it in order to generate the associated performance measure. In other cases, the data you need may have been collected by another department. For instance, in example 2, the delivery date was probably collected by the shipping and receiving department. Examine the data you need and determine if your own work group, department, or an external group is already collecting them.

If the data do not presently exist, the team will have to determine where to find it. The process of locating it is generally quite straightforward. This is particularly true if the team is measuring its own process. The measurement point is *usually* located at or near each critical activity identified in step 2. This is generally the case if your performance measure is measuring an activity within the process rather than the overall process itself. For performance measures that assess some overall aspect of a process, the collection point usually occurs at the culmination of a process.

More global performance measures generally require data from many sources, as in example 3. Before proceeding, the team must determine where the data are located, or where in the process, at which point in time, and at which physical location the data will be collected.

Continuing with the procurement example, we would probably find something like the following take place.

> The team has determined the raw data they will need to construct their performance measure and now they must locate the data. In this example, the process is rather simple. The contract delivery date is recorded within the procurement department itself on several documents and within a database so that retrieval will be trivial. The second data element is recorded by the shipping and receiving department and is likewise simple to extract. All that remains to reconstruct the performance measure are the computations with the data.

Identify the indicator

By this point, the team has determined what raw data they require, where it is located, and where it will be collected. To proceed, they must determine how they will actually measure or collect what they need. An indicator is required to accomplish the measurement.

An indicator is a device or person that is able to detect (sense) the presence or absence of some phenomena and (generally) provide a reading of the intensity (amount) of that phenomena in a quantifiable form with the appropriate units of measure. The indicator is what or who will do the measuring or data collection for your measurement system.

Indicators take many forms depending on what they are designed to measure. For technical and manufacturing processes, there are indicators that can accurately measure length (micrometers), temperature (thermocouples), voltage (digital voltmeters or digitizers), and so on. For less technical processes there are databases, log books, time cards, and check sheets. In some cases, the indicator takes a measurement and a person records the results. In other cases, only a human is capable of "sensing" some phenomena, and some other device is used to record the result. Many inspection activities can only be performed by humans. There are also automated data collection systems or indicators that require no human intervention other than calibration or maintenance. Many manufacturing processes employ such indicators to detect, measure, and record the presence of nonstandard products.

Choosing an indicator usually involves asking simple questions about the measurement you hope to take.

1. What am I trying to measure; what kinds of data are they?
2. Where will I take the measurement; where are the data?
3. Am I simply trying to measure the presence or absence of some feature? (Was the order placed? Was the report delivered? Did the computer help desk solve the problem?)
4. Do I need to sense the degree or magnitude of some feature or count how many?
5. How accurate and precise must my measurements be?
6. Do the measurements occur at a particular point in time or space?

In most cases, the answer will be rather obvious, but the team should be prepared to give some thought to how they will measure and collect their data. For instance, the need for accuracy and/or precision may rule out certain indicators. If you rely on a human as an indicator, you must consider all the possible biases that are inherent in human indicators. Step 6 discusses biases and their potential solution. Replacing human indicators with technological instruments may be the best solution if a particularly critical measurement requires unusual accuracy or precision.

When the team completes this step, they should have an indicator identified for each raw data element and should have determined where the indicator will be deployed.

> The procurement team determine that the indicator for their first data element (contract delivery date) will be the *buyer's diary*, an electronic database maintained by each buyer in company-supported software. The second indicator is determined to be the *receiving log*,

which is maintained at the receiving dock by a receiving clerk. This indicator provides the actual delivery date for each order. Having identified the indicators, the team can now acquire the necessary data.

Determine how often to take measurements

In this last activity, the team will determine how often measurements should be made. In a sense, there are two distinct types of measures taken when a performance measurement system is adopted. One type is the performance measure itself. This is generally taken (calculated) and reported over some regular or repeating time interval. In the procurement example, the performance measure is calculated and presumably reported at a frequency of once a month. Some performance measures are used to observe real-time trends in a process and may be measured and plotted daily. In general, the frequency of measurement for the performance measure is usually determined when the performance measure itself is determined. Often the unit of measure chosen as the performance measure contains or alludes to the frequency of measurement.

The other measure that should be addressed is that of the raw data themselves. The frequency with which raw data are collected or measured may have a significant impact on the interpretation of the performance measure. For some performance measures, this amounts to asking *how many* data are needed to make the measure valid or statistically significant. Each team or manager will have to determine how often measurements must be made (data taken) to ensure statistical significance and believable results.

Again, using the procurement example, the raw data for this measure are each time a buyer enters contract data into the database and each time the receiving clerk logs a delivery. It could be said, then, that the frequency of measurement or data collection is continuous; that is, data are rewarded each time a transaction or delivery occurs.

Processes that are repeated numerous times per hour may only require a sample measure of every tenth event or so. Other events, like the procurement example, are measured or recorded each time they happen. Teams should use their best judgment in choosing the frequency of data collection and should consult the company's statistician or quality consultant if there are questions.

Step 5: Identify responsible party/parties

Steps 1 through 4 are primarily team activities. To continue the performance measurement process, the responsible worker(s) and the responsible decision maker must be defined. (In some instances, one person may be responsible for the entire system.) It is now appropriate to determine who should

- Collect the data
- Analyze/report actual performance
- Compare actual performance with goal/standard
- Determine if corrective action is necessary
- Make changes

Ideally, responsibility should be assigned to individuals commensurate with authority. This means that each responsible party should

1. Know what the goals are
2. Know what the actual performance is
3. Have the authority to implement changes if performance does not conform to goals and standards

To hold someone responsible in the absence of authority prevents them from performing their job and creates the risk of unwarranted blame.

Step 6: Collect data

The determination of conformance depends on meaningful and valid data. Before you start out to collect a lot of new data, it is always wise to look at the data you already have to make certain you have extracted all the information you can from them. In addition, you may wish to refer back to step 2 and review the plan for good data collection.

Information, for a team, comprises the answers to your questions. Data are a set of facts presented in quantitative or descriptive form. Obviously, data must be specific enough to provide you with relevant information. There are two basic kinds of data.

- *Measured or variables data*: Data that may take on any value within some range. This type of data provides a more detailed history of your business process. This involves collecting numeric values that quantify a measurement and therefore require small samples. If the data set is potentially large, consider recording a representative sample for this type of data.

  Examples:
  - Cost of overnight mail
  - Dollar value of stock
  - Number of days it takes to solve a problem
  - Diameter of a shaft
  - Number of hours to process an engineering change request
  - Number of errors on a letter

- *Counted or attribute data*: Data that may take on only discrete values. Attribute data need not be numeric. These kinds of data are counted, not measured, and generally require large sample sizes to be useful. Counting methods include defective/nondefective, yes/no, and accept/reject.
  Examples:
  - Was the letter typed with no errors?
  - Did the meeting start on time?
  - Was the phone answered by the second ring?
  - Was the report turned in on schedule?

A system owner needs to supervise the data collection process to determine whether the data are being collected properly and whether people are carrying out their assignments. Some form of preliminary analysis is necessary during the data collection process. Is your measurement system functioning as designed? Check the frequency of data collection. Is it often enough? Is it too often? Make adjustments as necessary and provide feedback to the data collectors.

Data collection forms

There are two types of forms commonly used to aid in data collection, often in combination.

- *Check sheet*: A form specially designed so that the results can be readily interpreted from the form itself. This form of data collection is ideal for capturing special (worker-controlled) causes of process variation since the worker can interpret the results and take corrective actions immediately.
- *Data sheet*: A form designed to collect data in a simple tabular or columnar format (often related to time-dependent data). Specific bits of data—numbers, words, or marks—are entered in spaces on the sheet. As a result, additional processing is typically required after the data are collected in order to construct the tools needed for analysis. This form of data collection is usually used for capturing common (manager-controlled) causes of process variations.

Data collection system

This system ensures that all of our measurements are collected and stored. The type of data and frequency of measurement will help you determine how to collect it. Some data fit well into check sheets or data sheets that collect information in simple tabular or columnar formats.

Other measurements lend themselves to easy entry into a computer database. Whatever system is chosen should provide easy access and be understandable by those who are tasked with reviewing the data. Those tasked with performing the data collection should understand the data collection system, have the necessary forms at hand, be trained in the data collection, and have access to instructions pertaining to the system.

The data collected needs to be accurate. Inaccurate data may give the wrong answer to our information questions. One of the most troublesome sources of error is called bias. It is important to understand bias and to allow for this during the development and implementation of any data collection system. The design of data collection forms and processes can reduce bias.

Some types of biases that may occur are as follows:

- *Exclusion*: Some part of the process or the data has been left out of the data collection process.
- *Interaction*: The data collection itself interferes with the process it is measuring.
- *Perception*: The data collector biases (distorts) the data.
- *Operational*: The data collection procedures were not followed or were specified incorrectly or ambiguously.
- *Nonresponse*: Some of the data are missing or not obtained.
- *Estimation*: Statistical biases have been introduced.
- *Collection time period*: The time period or frequency selected for data collection distorts the data, typically by missing significant events or cyclic occurrences.

Step 7: Analyze/report actual performance

Before drawing conclusions from the data, you should verify that the data collection process has met the following requirements:

- Review the information questions that were originally asked. Do the collected data still appear to answer those questions?
- Is there any evidence of bias in the collection process?
- Is the number of observations collected the number specified? If not, why not?
- Do you have enough data to draw meaningful conclusions?

Once the raw data are collected are verified, it is time for analysis. In most instances, your recorded data are necessarily the actual performance measurement. Performance measurements are usually formulated based on one or more raw data inputs. Therefore, you need to assemble the raw data into a performance measurement.

The next step in analyzing data is deciding how you are going to present or display the data. You usually group the data in a form that makes it easier to draw conclusions. This grouping or summarizing may take several forms: tabulation, graphs, or statistical comparisons. Sometimes, single data grouping will suffice for the purpose of decision making. In more complex cases, and especially where larger amounts of data must be dealt with, multiple groupings are essential for creating a clear base for analysis.

After summarizing your data, develop your report. A number of tools are available to assist you. The following are some of the more widely used tools and concepts to help you in your reporting.

- Use spreadsheets and databases as appropriate to organize and categorize the data and to graphically show the trends. This will greatly improve the ease and quality of interpretation. Some of the more common graphic presentations are histograms, bar charts, pie charts, scatter diagrams, and control charts.
- Make the report comparative to the goals.
- Make use of summaries. The common purpose is to present a single important total rather than many subtotals. Through this summary, the reader is able to understand enough to judge whether to go into detail or to skip to the next summary.
- Be aware of pitfalls in your data presentation. Averaging your data on a monthly basis might shorten the amount of information presented but could hide variations within the monthly period. Choices of scales on graphs and plots could skew interpretation.
- Standardize the calendar so that the month begins and ends uniformly for all reports. Failing this, the relation of cause to effect is influenced by the fact that events tend to congest at the end of the reporting period.
- Adopt a standard format. Use the same size for all sheets or charts. As far as possible, use the same scales and headings.

Reports may take many forms. However, at this stage, the report is intended to be a status transfer of information to the responsible decision maker for the process. Therefore, the report will likely consist of sets of tables or charts that track the performance measures, supplemented with basic conclusions.

Step 8: Compare actual performance with goals/standards

Within their span of control, responsible workers compare actual performance with the goal or standard. If variance warrants action, a report is made to the responsible decision maker.

Once the comparison against the goal or standard is initially established, you have several alternatives available for possible actions. You can decide to

- Forget it; variance is not significant.
- Fix it (steps 9 and 10).
- Challenge the goal or standard (step 11).

If there is no significant variance, then continue the data collection cycle. If there is a variance between the goal and the performance measure, look at the magnitude. If it is significant, report to the decision maker. If a decision to implement a corrective action is warranted, go to step 9.

Step 9: Determine whether corrective action is necessary

Step 9 is a decision step. You can either change the process or change the goal. If the variance is large, you may have a problem with your process and will need to make corrections to bring the performance back in line with the desired goal or standard. To address these potential problems, you can form a quality improvement team or do a root cause analysis to evaluate. Consider, too, that the goal may have been unrealistic.

If the variance is small, your process is probably in good shape. But, you should consider reevaluating your goals to make them more challenging. In addition, if you do make changes to the process, you will need to reevaluate goals to make sure they are still viable.

1. Remove defects; in many cases this is worker controllable.
2. Remove the cause of defects. Depending on the defect cause, this may be worker or management controllable.
3. Attain a new state of process performance, one that will prevent defects from happening.
4. Maintain or enhance the efficiency and effectiveness of the process. This is an essential condition for continuing process improvement and ultimately increasing the competitiveness and profitability of the business itself.

Step 10: Make changes to bring process back in line with goals and standards

This is the final step in closing the feedback loop: making changes to bring the process back in line with the goal or standard. Changes comprise a number of actions that are carried out to achieve one or more of the correction objectives listed in step 9.

The prime result of these corrective actions should be the removal of all identified causes of defects, resulting in an improved or a new process.

Step 11: Determine if new goals or measures are needed

The decision to create new performance measures or goals will depend on three major factors.

1. The degree of success in achieving previous objectives
2. The extent of any change to the scope of the work processes
3. The adequacy of current measures to communicate improvement status relative to critical work processes

Goals need to be challenging but also realistically achievable. If previously set objectives were attained with great difficulty, or not reached at all, then it may be reasonable to readjust expectations. This also applies to the objectives that were too easily met. Extensive scope changes to the work processes will also necessitate establishing new performance measures and goals. Changes in performance measures and goals should be considered annually and integrated into planning and budgeting activities.

Important terms in performance measurement

- *Accuracy*: The closeness of a measurement to the accepted true value. The smaller the difference between the measurement and the true value, the more accurate the measurement.
- *Attribute data*: Data that may take on only discrete values; they need not be numeric. These kinds of data are counted, not measured, and generally require large sample sizes to be useful.
- *Bias (of measurement)*: A tendency or inclination of outlook that is a troublesome source of error in human sensing.
- *Check sheet*: A form specially designed so that results can be readily interpreted from the form itself.
- *Continuous improvement*: The ongoing improvement of products, services, and processes through incremental and measureable enhancements.
- *Control*: The set of activities employed to detect and correct variation in order to maintain or restore a desired state of conformance with quality goals.
- *Corrective action*: Measures taken to rectify conditions adverse to quality and, where necessary, to preclude repetition.
- *Critical activity*: Activity/activities that significantly impact total process efficiency, effectiveness, quality, timeliness, productivity, or

safety. At the management level, they impact management priorities, organizational goals, and external customer goals.

- *Customer*: An entity that receives products, services, or deliverables. Customers may be either internal or external.
- *Data*: Information or a set of facts presented in descriptive form. There are two basic kinds of data: *measured* (also known as *variables data*) and *counted* (also known as *attribute data*).
- *Data collection system*: A broadly defined term indicating sets of equipment, log books, data sheets, and personnel used to record and store the information required to generate the performance measurement of a process.
- *Data sheet*: A form designed to collect data in a simple tabular or columnar format. Specific bits of data—numbers, words, or marks—are entered into spaces on the sheet. Additional processing is typically required after the data are collected in order to construct the tools needed for analysis.
- *Defect*: A nonconformance to the product quality goals; this leads to customer dissatisfaction.
- *Effectiveness*: A process characteristic indicating the degree to which the process output (work product) conforms to requirement.
- *Efficiency*: A process characteristic indicating the degree to which the process produces the required output at minimum cost.
- *Feedback*: Communication of quality performance to sources that can take appropriate action.
- *Feedback loop*: A systematic series of steps for maintaining conformance to quality goals by feeding back performance data for evaluation and corrective action. This is the basic mechanism for quality control.
- *Frequency*: One of the components of a performance measurement that indicates how often the measurement is made.
- *Goal*: A proposed statement of attainment/achievement, with the implication of sustained effort and energy.
- *Indicator*: A specialized detecting device designed to recognize the presence and intensity of certain phenomena and to convert this sensed knowledge into information.
- *Management assessment*: The determination of the appropriateness, thoroughness, and effectiveness of management processes.
- *Optimum*: A planned result that meets the needs of customer and supplier alike, meets competition, and minimizes the customer's and supplier's combined costs.
- *Organization*: Any program, facility, operation, or division.
- *Performance measure*: A generic term encompassing the quantitative basis by which objectives are established and performance is assessed and gauged. Performance measures include performance

objectives and criteria (POCs), performance indicators, and any other means of evaluating the success in achieving a specified goal.
- *Performance measurement category*: An organizationally dependent grouping of related performance measures that convey a characteristic of a process, such as cycle time.
- *Performance measurement system*: The organized means of defining, collecting, analyzing, reporting, and making decisions regarding all performance measures within a process.
- *Precision*: The closeness of a group of repeated measurements to their mean value. The smaller the difference between the group of repeat measurements and the mean value, the more precise the instrument. Precision is an indicator of the repeatability, or consistency, of the measurement.
- *Process*: Any activity or group of activities that takes an input, adds value to it, and provides an output to a customer. The logical organization of people, materials, energy, equipment, and procedures into work activities designed to produce a specified end result (work product).
- *Productivity*: The value added by the process divided by the value of the labor and capital consumed.
- *Quality*: The degree to which a product or service meets customer requirements and expectations.
- *Raw data*: Data not processed or interpreted.
- *Safety*: Measures the overall health of the organization and the working environment of its employee.
- *Self-assessment*: The continuous process of comparing performance with desired objectives to identify opportunities for improvement. Assessments conducted by individuals, groups, or organizations relating to their own work.
- *Timeliness*: Measures whether a unit of work was done correctly and on time. Criteria must be established to define what constitutes timeliness for a given unit of work. The criterion is usually based on customer requirements.
- *Unit of measure*: A defined amount of some quality feature that permits evaluation of that feature in numbers.
- *Validation*: A determination that an improvement action is functioning as designed and has eliminated the specific issue for which it was designed.
- *Variable data*: Data that may take on any value within some range. It provides a more detailed history of a business process. This involves collecting numeric values that quantify a measurement and therefore requires small samples.
- *Variance*: In quality management terminology, any nonconformance to specification.

- *Verification*: The determination that an improvement action has been implemented as designed.
- *Worker controllable*: A state in which the worker possesses (1) the means of knowing what the quality goal is, (2) the means of knowing what the actual quality performance is, and (3) the means of changing performance in the event of nonconformance.

In concluding this chapter, it is important to note that, along with performance measures, the broad field of ergonomics is essential for a sustainable work design for occupational advancement (Konz and Johnson, 2004; Peacock, 2009).

References

Badiru, A. B. (1994), Multifactor learning and forgetting models for productivity and performance analysis, *International Journal of Human Factors in Manufacturing*, Vol. 4, No. 1, pp. 37–54.

Badiru, A. B. and A. Ijaduola (2009), Half-life theory of learning curves for system performance analysis, *IEEE Systems Journal*, Vol. 3, No. 2, June, pp. 154–165.

Badiru, I. A. (2016), Comments about work management, interview with an auto industry senior engineer about corporate views of work design, Beavercreek, OH, October 29.

Badiru, A. B. and B. J. Ayeni (1993), *Practitioner's Guide to Quality and Process Improvement*, London, Chapman & Hall.

DOE (1995), How to measure performance: A handbook of techniques and tools, US Department of Energy, Defense Programs, Special Projects Group (DP-31), Washington, DC.

Geisler, J. (2012), *Work Happy: What Great Bosses Know*, New York, Hachette.

Harrington, H. J. (1991), *Business Process Improvement*, New York, McGraw-Hill.

Juran, J. M. (1964), *Managerial Breakthrough*, New York, McGraw-Hill.

Juran, J. M. (1988), *Juran on Planning for Quality*, New York, The Free Press.

Juran, J. M. (1989), *Juran on Leadership for Quality*, New York, The Free Press.

Juran, J. M. and F. M. Gryna, Jr (1980), *Quality Planning and Analysis*, New York, McGraw-Hill.

Juran Institute (1989), *Quality Improvement Tools*, Wilton, CT, Juran Institute.

Juran Institute (1990), *Business Process Quality Management*, Wilton, CT, Juran Institute.

Konz, S. and S. Johnson (2004), *Work Design: Occupational Ergonomics*, 6th Edition, Scottsdale, AZ, Holcomb Hathaway Publishers.

Pall, G. A. (1987), *Quality Process Management*, Englewood Cliffs, NJ, Prentice-Hall.

Peacock, B. (2009), *The Laws and Rules of Ergonomics in Design*, Singapore, Eazi Printing Ltd (www.eazi.sg/printing/).

Sieger, D. B. and A. B. Badiru (1995), A performance parameter model for design integration, *Engineering Design and Automation*, Vol. 1, No. 3, pp. 137–148.

chapter five

Cognitive task evaluation

Cognitive task evaluation plays and important role in work system design, so engineers and analysts can understand behaviors that cause errors. In manufacturing, this is becoming more important as lean production systems require more cognitive work. This chapter presents some cognitive task analysis methods for process design and modification to aid in improving work efficiency.

Cognitive ergonomics

According to Karwowski (2012), ergonomics "promotes a holistic, human-centered approach to work systems design that considers the physical, cognitive, social, organizational, environmental and other relevant factors." Coelho (2011) describes ergonomics as the scientific field that involves the interactions between humans and other components of a system, and says that this profession applies theoretical principles, data, and methods to designing an optimal system for the well-being of the human and for performance. Hollnagel (1997) states the principle meaning of ergonomics is the science or study of work.

The objective of ergonomics is to understand the interactions of work systems and their operators in order to enhance performance, increase safety, and improve user satisfaction (Bouargane and Cherkaoui, 2015). Designing systems with a better adaptation of work for the human operator can result in greater efficiency and satisfaction for the operator, as well as improved productivity, reduced accidents, and lower turnover for the company (Tajri and Cherkaoui, 2013). There are three domains within the ergonomic discipline: physical, cognitive, and organizational. This chapter will focus on the cognitive demands of an operator's performance.

The term *cognitive ergonomics* was coined in 1975 by Sime, Fitter, and Greene (Green and Hoc, 1991). Cognitive ergonomics involves the relationship between tools and their users, putting emphasis on the cognitive processes for understanding, reasoning, and usage of knowledge (Green and Hoc, 1991). According to the International Ergonomic Association, cognitive ergonomics relates to mental processes such as perception, memory, reasoning, and motor response, as they influence the interactions between humans and other elements of a work system; it is the ergonomics of mental processes for improving operator performance. Cognitive ergonomics involves the psychological aspects of work, evaluating how work affects the mind and how the mind affects work (Hollnagel, 1997). It addresses how people process information, generate decisions, and perform actions (Belkic and Savic, 2008). As stated by the International Ergonomic Association (2016), relative to human–system design, some applicable topics in cognitive ergonomics are

- Mental workload
- Decision making
- Skilled performance
- Work stress
- Human–computer interaction
- Human reliability

Over the past 40 years, high-speed technological advances have propelled cognitive ergonomics development (Hoc, 2008). According to Kramer (2009), the concept of cognitive ergonomics emphasizes attaining a balance between the human's cognitive abilities and limitations, in addition to the machine, task, and environment. This can lead to a reduction in errors and better performance while working, even in repetitive physical tasks. Cognitive ergonomics emphasizes the quality of work, including the outcome, unlike traditional or physical ergonomics, which looks at the quality of working (Hollnagel, 1997). Traditional ergonomics aims at reducing operator fatigue and discomfort in an effort to improve throughput. However, the nature of human work has transformed considerably from working with the body to increased cognitive requirements as industrial systems have become more automated. The expanded role of technology, along with the use of complex procedures, has imposed more demands on operators (Stanton et al., 2010). Instead of physical endurance and physical strength, there is an increased need for sustained attention, problem solving, and reasoning. In addition, physically demanding work that is performed concurrently with a cognitive task can impact mental workload by weakening mental processing or decreasing performance (DiDomenico and Nussbaum, 2011). Therefore, the use of cognitive ergonomics plays an important role in optimizing operator and system performance (Zhang et al., 2014). In order to improve human–system design for performance, cognitive ergonomics should be included in work system design, in order to accomplish an in-depth evaluation of the cognitive elements required to perform a task. There are numerous cognitive task analysis (CTA) techniques for evaluating the cognitive processes of work systems, including

- Applied cognitive task analysis (Militello and Hutton, 1998)
- Cognitive work analysis (Vicente, 1999)

- The critical decision method (Klein et al., 1989)
- Cognitive walkthrough (Polson et al., 1992)

The concept of CTA uses a set of techniques designed to extract information relative to the knowledge, thought processes, and goal structures that underlie observable performance to describe and represent the cognitive elements (Antonova and Stefanov, 2011). Analysts use CTA to collect accurate and comprehensive descriptions of cognitive processes and decisions (Clark et al., 2008). According to Antonova and Stefanov (2011), CTA uses the three broad families of knowledge elicitation: observation and interviews, process tracing, and conceptual techniques. Cooke (1994) describes the categories as follows:

1. Observations and interviews involve the informal process of watching the experts and talking with them.
2. Process tracing focuses on the cognitive structure of a specific task while it is being performed by an expert.
3. Conceptual techniques take data from multiple experts and produce a structured representation of the domain concepts.

There are five common steps in most of the major CTA methods (Clark et al., 2008); the steps and order of execution are illustrated in Figure 5.1.

Clark et al. (2008) describe the five common steps of CTA methods as follows:

- *Step 1*, the collection of preliminary knowledge, identifies the sequence of work elements that will be analyzed. In addition, the analyst identifies the experts that will participate in the knowledge elicitation stage during step 3.
- *Step 2* identifies knowledge representations. This step uses the information collected in step 1 to identify subtasks and the knowledge required to accomplish the task. Some approaches that can be used for knowledge representation are concept maps, flow charts, and a learning hierarchy.
- *Step 3* applies the focused knowledge elicitation methods previously discussed. Recall, during knowledge elicitation, that the analyst employs various techniques to gather the knowledge obtained in the preceding step. Analysts typically choose methods suitable for the knowledge type, as established by the knowledge representations pinpointed for each task; thus, most knowledge elicitation efforts involve multiple techniques.
- *Step 4* analyzes and verifies the data acquired, because many knowledge elicitation techniques are less formal and need the analyst to code and format the results for the verification, validation, and

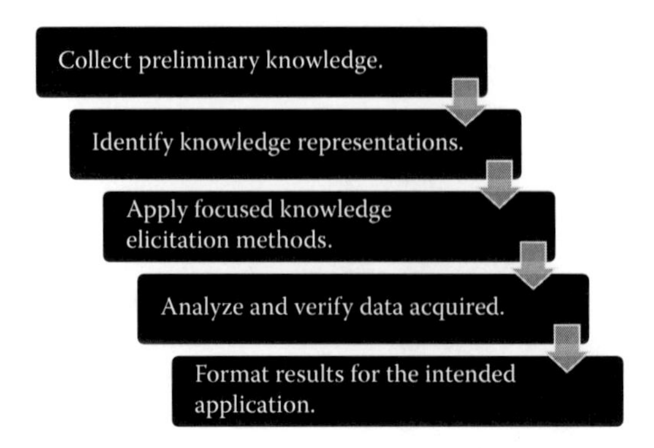

Figure 5.1 Five common steps of CTA methods.

applicability of usage into the intended application. After the coding process, the structured output is given to the participating subject matter experts (SMEs) for verification, modification, and revision to confirm that the representations of tasks and their underlying cognitive elements are complete and accurate. Once the information in the structured output has been verified or revised by the SMEs, the analyst should then compare it with the output of the other SMEs to validate that the results accurately indicate the desired knowledge representation.

- *Step 5* formats the results for the intended application. For many of the CTA methods, the results must be transformed into cognitive models that indicate the underlying skills, mental models, and problem-solving strategies used by the SMEs when executing complex tasks.

This chapter will review two CTA methods for the evaluation of cognitive elements in work design: cognitive work analysis (CWA) and applied cognitive task analysis (ACTA). CWA techniques are considered complex and require a substantial amount of training. However, ACTA is regarded as easy to use and does not necessitate cognitive training.

Cognitive work analysis

CWA is a structured approach that systematically shifts from analyzing mental demands to determining visualizations and the decision-aiding concepts that can assist the understanding of cognitive work (Stanton et al., 2004). The CWA framework (Vicente, 1999) was originally developed at Riso National Laboratories in Denmark by Jens Rasmussen and his colleagues to analyze cognitive work (Rasmussen et al., 1994). CWA is a method used to analyze, design, and evaluate interactive systems

(Sanderson, 1998). According to Stanton et al. (2010), this approach can be applied to define

1. The functional properties of the work domain to be evaluated
2. The nature of the tasks that are worked within the system
3. The roles of the different actors that exist within the system
4. The cognitive skills and strategies that they utilize to accomplish activity within the system

CWA provides tools for characterizing, comparing, and evaluating different designs during the development phase of a system, and it can be a very effective tool for evaluating different design proposals (Sanderson et al., 1999). CWA methods have been applied in guiding the redesign of a human–system interface for a newly automated system, and in the development of an information system or decision aid to support a set of tasks where such support did not formerly exist (Bisantz et al., 2003). Table 5.1 outlines a variety of domains in which CWA has been applied.

Figure 5.2 lays out the five phases of CWA, which model different constraint sets: work domain analysis (WDA), control task analysis, strategies analysis, social organization and cooperation analysis (SOCA), and worker competencies analysis.

The WDA phase models the system, considering its purposes and the limitations enforced by the environment. WDA provides a framework for developing requirements by evaluating why a new system should exist, what its environmental context will be, and what functions the new system should employ (Sanderson et al., 1999). Control task analysis is utilized to define the tasks within the system and the constraints inflicted on these tasks under various conditions. Strategies analysis is employed to see how different activities can be accomplished. In order to determine

Table 5.1 CWA applications

Authors	Year of publication	Domain
Ashoori et al.	2014	Health care
Bisantz et al.	2003	Military
Burns et al.	2008	Health care
Gous	2013	Air defense
Higgins	1998	Manufacturing
Lintern and Naikar	1998	Training
McIlroy and Stanton	2015	Ecological interface design
Nirula and Woodruff	2006	Education
Qureshi	2000	Air defense
Sanderson et al.	1999	Air defense
Vicente	1999	Nuclear power

Figure 5.2 Phases of CWA.

how social and technical factors can work in unison to enhance system performance, SOCA is used to determine how activity and its related strategies can be allocated to human operators. This phase simply studies who carries out the work elements and how the work is allocated. Worker competencies analysis pinpoints the capabilities that an ideal worker should demonstrate by concentrating on the strengths and weaknesses of the human operator relative to the work system design. Not all the phases of CWA must be performed; the technique may be regarded as a toolkit from which the analyst may use single or multiple parts to suit his or her needs (McIlory and Stanton, 2015). The overall purpose of this analysis is to improve work efficiency by analyzing the design of work using information flow and representation, task specification, operator training, and skills requirements, along with the social structures of work teams (Allwood et al., 2016).

Applied cognitive task analysis

ACTA offers techniques that can be employed to analyze the mental demands involved in a task. These techniques were developed as part of a project funded by the Navy Personnel Research and Development Center (Militello and Hutton, 1998). The ACTA methodology, compared with traditional task analysis, is most popular among system designers in exploring complex and dynamic domains (Prasanna et al., 2009). It is appropriate for job domains where observational data are difficult to acquire, and can be applied for the identification and classification of complex skills (Antonova and Stefanov, 2011). ACTA consists of three interview methods that assist the analyst in acquiring information about the skills and cognitive demands required for a task.

- Task diagram interview
- Knowledge audit interview
- Simulation interview

ACTA is a flexible, easy-to-use method that doesn't require the analyst to be an expert in the knowledge domain (Antonova and Stefanov, 2011). As shown in Figure 5.3, there are seven steps to the ACTA process. In addition to operator interviews, the ACTA process uses methods such as task walkthroughs and observations to collect data.

Step 1 is to select and define the work elements of the task to be analyzed. Next, in step 2, the correct set of SMEs should be identified. During step 3, the task should be observed by the analyst in order to follow the participant roles throughout the task. Step 4 is the task diagram interview, which provides the analyst with an overview of the task that is being studied. This interview offers a high-level review of the cognitive elements in the work task; it limits the SME to between three and six steps for describing the work elements to ensure that time is not wasted probing into miniscule details during this high-level interview (Militello et al., 1997). During this process, the tasks that are extremely challenging are highlighted. A diagram should be produced to illustrate the task steps that necessitate the most cognitive skills. In step 5, the knowledge audit interview draws attention to the tasks that entail expertise. The SME is questioned for examples within the context of the task, critical cues, and decision-making strategies. The output of the knowledge audit interview is a table that provides insight as to why the situations that require expertise present a challenge for less-experienced operators. This table includes examples of situations in which experience has to be called into play, along with the cues and strategies used in dealing with difficult situations (Militello et al., 1997). Next, the simulation interview (step 6)

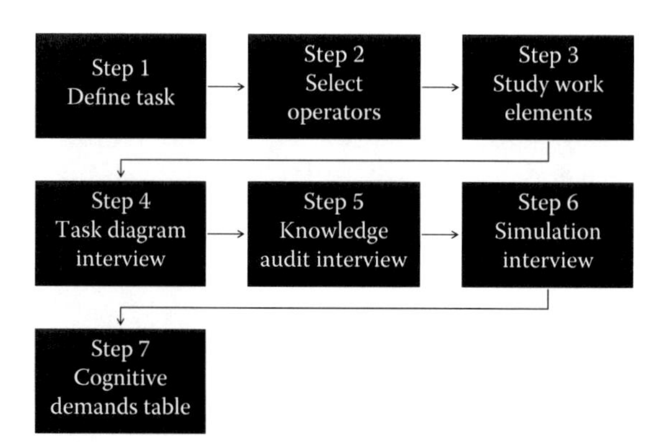

Figure 5.3 ACTA procedural steps.

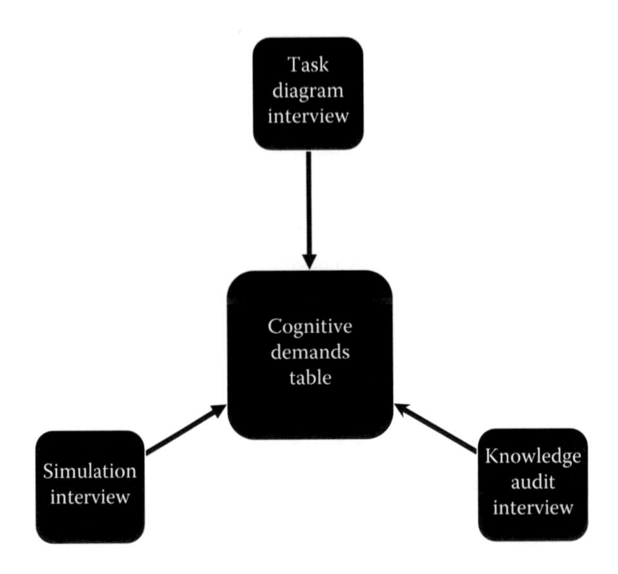

Figure 5.4 Data integration for cognitive demands table development. (Adapted from R. Prasanna et al., *Proceedings of the 6th International ISCRAM Conference*, 1–10, 2009.)

takes place while the task under analysis is being performed. The SME is probed on the cognitive processes relative to issues such as situation awareness and potential errors. It allows the analyst to understand the cognitive processes used within the context of an incident (Militello and Hutton, 1998). Lastly (step 7), the cognitive demands table integrates the data obtained from the task diagram, knowledge audit, and simulation interviews into one organized format.

Figure 5.4 illustrates the integration of the data collection for developing the cognitive demands table. Not all information discussed in the interviews will be applicable to the goals of the task evaluation. Therefore, the cognitive demands table is intended to provide a format for the analyst in order to focus the analysis of the data collection on the project goals. This format assists the analyst with identifying common themes in the data, as well as conflicting information given by various SMEs (Militello et al., 1997).

Summary

Cognitive ergonomics plays an important role in the design of work systems. The trends of research in ergonomics are shifting from the physical and perceptual to the cognitive aspects of tasks, product design, and human interface design, because this approach has proven to improve work efficiency and decrease human errors (Murata, 2000). The goal of cognitive ergonomists is to design tools and systems that support human mental processes and minimize biases, based on the scientific understanding of attention, perception, memory, mental models. and mental workload in an effort to expand human cognitive capabilities (Bouargane and Cherkaoui, 2015). This concept of cognitive ergonomics can assist with understanding the types of behaviors that cause errors, so systems can be designed to reduce the frequency of these errors and their outcomes. There are many cognitive ergonomic techniques that can be utilized for collecting the appropriate system information, in order to present the data in the right format such that effective analysis can be completed on the cognitive requirements for a successful process design and/or modification. This is becoming increasingly more important in manufacturing work design as many production facilities are moving from mass to lean production environments. In lean production environments, the duties of operators involve a higher level of cognitive work in comparison with the duties of workers in a mass production environment (Genaidy and Karwowski, 2003).

References

Allwood, J. M., Childs, T. H. C., Clare, A. T., Silva, A. K. M. D., Dhokia, V., Hutchings, I. M., et al. (2016). Manufacturing at double the speed. *Journal of Materials Processing Technology*, *229*, 729–757.

Antonova, A., and Stefanov, K. (2011). Applied cognitive task analysis in the context of serious games development. *Proceedings of the Third International Conference on Software, Services and Semantic Technologies (S3T) 2011*, pp. 175–182.

Ashoori, M., Burns, C. M., d'Entremont, B., and Momtahan, K. (2014). Using team cognitive work analysis to reveal healthcare team interactions in a birthing unit. *Ergonomics*, *57*(7), 973–986.

Belkic, K., and Savic, C. (2008). The occupational stress index: An approach derived from cognitive ergonomics applicable to clinical practice. *Scandinavian Journal of Work, Environment & Health*, *34*(6), 169.

Bisantz, A. M., Roth, E., Brickman, B., Gosbee, L. L., Hettinger, L., and McKinney, J. (2003). Integrating cognitive analyses in a large-scale system design process. *International Journal of Human–Computer Studies*, *58*(2), 177–206.

Bouargane, L., and Cherkaoui, A. (2015). Towards an explicative model of human cognitive process in a hidden hazardous situation and a cognitive ergonomics intervention in railway environment. *Proceedings of the International Conference on Industrial Engineering and Systems Management (IESM) 2015*, pp. 968–976.

Burns, C. M., Enomoto, Y., and Momtahan, K. (2008). A cognitive work analysis of cardiac care nurses performing teletriage. In A. M. Bisantz and C. M. Burns (eds.), *Applications of Cognitive Work Analysis*, pp. 149–174. Boca Raton, FL: CRC Press.

Clark, R. E., Feldon, D., van Merriënboer, J. J., Yates, K., and Early, S. (2008). Cognitive task analysis. *Handbook of Research on Educational Communications and Technology*, *3*, 577–593.

Coelho, D. A. (2011). Human factors and ergonomics: A growing discipline with multiple goals. Highlights - The Quarterly Newsletter, *International Journal of Human Factors and Ergonomics*, Winter ed. http://www.inderscience.com/info/highlights/2011/winter_ijhfe.pdf (accessed February 2017).

Cooke, N. J. (1994). Varieties of knowledge elicitation techniques. *International Journal of Human–Computer Studies*, *41*(6), 801–849.

DiDomenico, A., and Nussbaum, M. A. (2011). Effects of different physical workload parameters on mental workload and performance. *International Journal of Industrial Ergonomics*, *41*(3), 255–260.

Genaidy, A. M., and Karwowski, W. (2003). Human performance in lean production environment: Critical assessment and research framework. *Human Factors and Ergonomics in Manufacturing & Service Industries*, *13*(4), 317–330.

Gous, E. (2013). Utilising cognitive work analysis for the design and evaluation of command and control user interfaces. *Proceedings of the International Conference on Adaptive Science and Technology 2013*, pp. 1–7.

Green, T. R., and Hoc, J. (1991). What is cognitive ergonomics? *Le Travail Humain*, *54*, 291–304.

Higgins, P. G. (1998). Extending cognitive work analysis to manufacturing scheduling. *Proceedings of the Australasian Computer–Human Interaction Conference 1998*, pp. 236–243.

Hoc, J. (2008). Cognitive ergonomics: A multidisciplinary venture. *Ergonomics*, *51*(1), 71–75.

Hollnagel, E. (1997). Cognitive ergonomics: It's all in the mind. *Ergonomics*, *40*(10), 1170–1182.

International Ergonomic Association. Ergonomics: Human centered design. http://www.iea.cc/whats/index.html (accessed September 2016).

Karwowski, W. (2012). The discipline of human factors and ergonomics. In G. Salvendy (ed.), *Handbook of Human Factors and Ergonomics* (4th ed.), pp. 3–37. Hoboken, NJ: Wiley.

Klein, G. A., Calderwood, R., and Macgregor, D. (1989). Critical decision method for eliciting knowledge. *IEEE Transactions on Systems, Man, and Cybernetics*, *19*(3), 462–472.

Kramer, A. (2009). Cognitive Ergonomics 101: Improving Mental Performance. http://old.askergoworks.com/news/18/Cognitive-Ergonomics-101-Improving-Mental-Performance.aspx (accessed February 2017).

Lintern, G., and Naikar, N. (1998). Cognitive work analysis for training system design. *Proceedings of the Australasian Computer–Human Interaction Conference 1998*, pp. 252–259.

McIlroy, R. C., and Stanton, N. A. (2015). Ecological interface design two decades on: Whatever happened to the SRK taxonomy? *IEEE Transactions on Human–Machine Systems*, *45*(2), 145–163.

Militello, L. G., and Hutton, R. J. (1998). Applied cognitive task analysis (ACTA): A practitioner's toolkit for understanding cognitive task demands. *Ergonomics*, *41*(11), 1618–1641.

Militello, L. G., Hutton, R. J., Pliske, R. M., Knight, B. J., and Klein, G. (1997). *Applied Cognitive Task Analysis (ACTA) Methodology*. Fairborn, OH: Klein.

Murata, A. (2000). Ergonomics and cognitive engineering for robot–human cooperation. *Proceedings of the 9th IEEE International Workshop on Robot and Human Interactive Communication (RO-MAN) 2000*, pp. 206–211.
Nirula, L., and Woodruff, E. (2006). Cognitive work analysis and design research: Designing for mobile human–technology interaction within elementary classrooms. *Proceedings of the Fourth IEEE International Workshop on Wireless, Mobile and Ubiquitous Technology in Education 2006 (WMTE '06)*, pp. 32–35.
Polson, P. G., Lewis, C., Rieman, J., and Wharton, C. (1992). Cognitive walkthroughs: A method for theory-based evaluation of user interfaces. *International Journal of Man–Machine Studies*, *36*(5), 741–773.
Prasanna, R., Yang, L., and King, M. (2009). GDIA: A cognitive task analysis protocol to capture the information requirements of emergency first responders. *Proceedings of the 6th International ISCRAM Conference*, pp. 1–10.
Qureshi, Z. H. (2000). Modelling decision-making in tactical airborne environments using cognitive work analysis-based techniques. *Proceedings of the 19th Digital Avionics Systems Conference (DASC) 2000*, Vol. 2, pp. 5B3/1-5B3/8.
Rasmussen, J., Pejtersen, A. M., and Goodstein, L. P. (1994). *Cognitive Systems Engineering*. New York: Wiley.
Sanderson, P. (1998). Cognitive work analysis and the analysis, design, and evaluation of human–computer interactive systems. *Proceedings of the Australasian Computer–Human Interaction Conference 1998*, pp. 220–227.
Sanderson, P., Naikar, N., Lintern, G., and Goss, S. (1999). Use of cognitive work analysis across the system life cycle: From requirements to decommissioning. *Proceedings of the Human Factors and Ergonomics Society Annual Meeting*, *43*(3), pp. 318–322.
Stanton, N. A., Hedge, A., Brookhuis, K., Salas, E., and Hendrick, H. W. (2004). *Handbook of Human Factors and Ergonomics Methods*. Boca Raton: CRC Press.
Stanton, N. A., Salmon, P. M., Walker, G. H., Baber, C., and Jenkins, D. P. (2010). *Human Factors Methods: A Practical Guide for Engineering and Design*, 1st Ed. Burlington, VT: Ashgate.
Tajri, I., and Cherkaoui, A. (2013). Proposal of a model to explain the cognitive process of employee, the occurrence of stress and cognitive ergonomics intervention in a total lean environment. *Proceedings of the International Conference on Industrial Engineering and Systems Management (IESM) 2013*, pp. 1–10.
Vicente, K. J. (1999). *Cognitive Work Analysis: Toward Safe, Productive, and Healthy Computer-Based Work*. Mahwah, NJ: CRC Press.
Zhang, H., Zhu, Y., Maniyeri, J., and Guan, C. (2014). Detection of variations in cognitive workload using multi-modality physiological sensors and a large margin unbiased regression machine. *Proceedings of the 36th Annual International Conference of the IEEE Engineering in Medicine and Biology Society*, pp. 2985–2988.

chapter six

Mental workload measures for work evaluation

This chapter covers mental workload (MWL) measures for work evaluation to assess levels of mental load imposed by a system. MWL is an area of study within cognitive ergonomics. MWL measures are major factors in system design and the analysis of occupational tasks (DiDomenico and Nussbaum, 2011). MWL assessments can be used to estimate performance requirements for the design of systems and to predict workload changes with various system modifications (May et al., 1990). Many modern systems put increasing demands on individuals by requiring them to perform physically demanding tasks simultaneously with more cognitive responsibilities. As these procedures for human–machine systems are becoming more complex and automated, the need for MWL assessment is becoming more critical (Rubio et al., 2004). In order to develop improved work systems that are globally competitive, the process designer should consider MWL evaluations for the development of optimal work performance and error reduction.

Information processing

Information processing is extremely important to human performance, because in situations where humans interact with work systems, "the operator must perceive information, must transform that information into different forms, must take actions on the basis of the perceived and transformed information, and must process the feedback from that action, assessing its effect on the environment" (Wickens and Carswell, 2006). This information is received visually or via an auditory organ (perception) and must be understood in order to interpret the meaning of the perceived information so that decisions can be made using the knowledge stored in the memory system (cognition) (Murata, 2000).

According to the Wickens and Hollands (2000) framework, sensory memory, working memory, and long-term memory are the three stages of information processing. Sensory memory is associated with interpreting data based on past experiences that were sensed or perceived. However, working memory (i.e., short-term memory) is a brain system that deals with the immediate, conscious processing of data; it is of limited capacity and heavily demanding of the attention, as it deals with information that is not sensed or perceived but is an internal thought. An example of working memory utilization is quality inspection (i.e., recalling memorized criteria for acceptable quality), which places a sustained load on the working memory. This demand may fatigue the operator or otherwise affect their performance in the task (Pesante et al., 2001). Long-term memory is the cognitive storage of data; a vast amount of information is kept, but it is not always possible to totally recall it. It is important that the operator is not oversaturated with tasks because mental overload could lead to performance degradation. MWL or simply *workload* is the interaction between an operator and his or her assigned task in a human–machine system. In simplistic terms, it is the demand on an operator.

Mental workload

MWL is defined as a multidimensional model that encompasses task and performance demands along with operator skill and attention (Stanton et al., 2004). MWL measures can offer awareness as to where increased task demands could lead to an adverse influence on human performance. The demands of a task may require completing physical activities and/or performing cognitive tasks. Understanding the task demands is important for work design, in order to design processes to reduce stress levels, accident rates, and errors. When the sum or the difficulty of the tasks required to complete a process increases, or when the time allocated for task completion decreases, workload generally rises (Colombi et al., 2012). Therefore, increased safety, health, comfort, and long-range productivity can be accomplished by regulating the task demands, such that the operator is neither underloaded nor overloaded (Rubio et al., 2004). This is relevant to work design, because high cognitive load for extended periods of time can lead to unproductive processes and poor performance as well as ergonomic problems and mental health symptoms (Lindblom and Thorvald, 2014).

MWL can affect an operator's ability to attain or maintain the desired performance levels (Xie and Salvendy, 2000). The concept of MWL deals with the difference between the amount of resources available to a person and the quantity of resources needed for the task (Jung and Jung, 2001). For this reason, high workload levels occur when the task demands exceed the operator's mental resource capacity (Loft et al., 2007). Therefore, MWL can be modified by altering the amount of resources available within the person or the demands required by the task on the operator. Multiple resource theory (MRT) is a model that can be employed to assess the various mental resources for various conditions in the design of the human–machine system.

According to Estes (2015), workload measurements became an area of significant research in the late 1960s, when a variety of techniques were being produced to measure them. Workload measurements have been utilized in various functions, including system desirability and system optimization. As the operator works the system, either with a simulation or the existing process, the operator's MWL can be evaluated and measured. MWL can be fragmented into two segments: effective and ineffective. Effective workload (i.e., task load) is the minimum amount of workload produced by the task requirements. The operator directly accomplishes the work with this portion of the workload (Xie and Salvendy, 2000). In contrast, ineffective workload should be avoided by workers because it does not directly contribute to completing a task (Xie and Salvendy, 2000). Ineffective workload can be lowered by learning and training. As operators' skill levels increase, their proficiency at completing a task increases and their perception of workload decreases (Colombi et al., 2012). When executing tasks, effective workload corresponds to fast and accurate actions, whereas ineffective workload is associated with error and inaccuracy. Effective workload necessitates the smallest amount of attention because it correlates to efficiency; however, "more attention is needed under the ineffective mode to control each stage of information processing" (Xie and Salvendy, 2000). As shown in Figure 6.1, there are three categories of MWL measures: subjective, physiological, and performance. These different measures assess the different dimensions of MWL, so it is advisable to use multiple measures for better diagnosticity (Miller, 2001; Tsang and Vidulich, 2006).

Subjective measures

Subjective workload measures involve the operator by including his or her judgment on the cognitive effort entailed to accomplish a task subsequent to completing it (Cegarra and Chevalier, 2008; Stanton et al., 2010; Hertzum and Holmegaard, 2013). These assessments are based on an individual's perception of the effort applied in performing a task. This method is typically not complicated to administer and has gained considerable theoretical support for its ability to provide sensitive measures on operator workload (Rubio et al., 2004). As cited by Bommer and Fendley (2015), subjective MWL measures have been widely accepted and have proven to be valid and reliable.

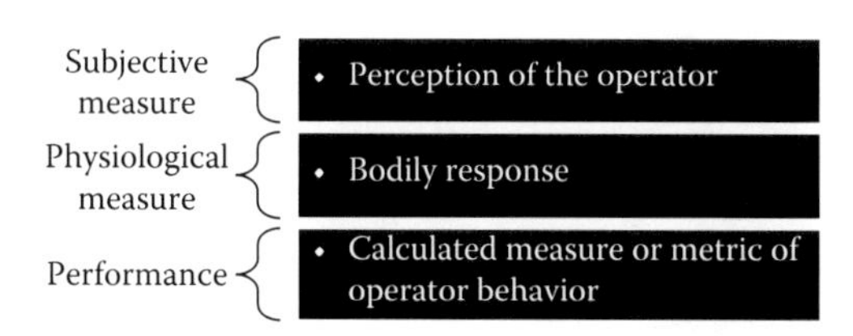

Figure 6.1 Measures for MWL analysis.

Subjective workload measures attempt to quantify the effort utilized while performing a task. They use numerical ratings that will not measure task performance or physiological responses associated directly with the work applied (DiDomenico and Nussbaum, 2008). This method, in general, is a good indicator of overall workload, because most of these techniques will not pinpoint or differentiate differences in loading peaks. There are a host of subjective methods to measure MWL that provide useful information, including

- Cooper–Harper scales (Cooper and Harper, 1969)
- The Bedford scale (Roscoe and Ellis, 1990)
- The subjective workload dominance technique (Vidullch et al., 1991)
- The subjective workload assessment technique (Reid and Nygren, 1988)
- The National Aeronautics and Space Administration Task Load Index (NASA-TLX) (Hart and Staveland, 1988)
- Workload profile (WP) (Tsang and Velazquez, 1996)

In this chapter, two subjective MWL methods will be discussed: the NASA-TLX and WP. These methods were chosen because the NASA-TLX has been commonly used across diverse domains and WP is based on the sound underpinning theory of MRT, which can aid in pinpointing the mental resource causing the workload.

National Aeronautics and Space Administration Task Load Index
The NASA-TLX is the most commonly applied MWL assessment approach of the subjective category. It has been utilized in various domains including civil and military aviation, driving, nuclear power plant control room operation, and air traffic control (Stanton et al., 2010), has been taught in university courses, and has been cited in previous studies, including Endsley and Rodgers (1997), Bertram et al. (1992), and Cao et al. (2009). Its application has expanded beyond its original applications in aviation and crew complement. It is "being used as a benchmark against which the efficacies of other measures, theories or models are judged" (Hart, 2006).

The NASA-TLX uses a multidimensional scale to quantify the task performance of the operator. Its multidimensional scale consists of six subscales.

- Mental demand
- Physical demand
- Temporal demand
- Effort
- Performance
- Frustration level

Mental demand measures the perceptual elements of the task including thinking, looking, and searching activities. Physical demand evaluates whether the physical activity was laborious or relaxed. Temporal demand measures how much pressure is created by the pace of the activity. Effort assesses how challenging it was to accomplish the tasks. The performance dimension takes into account how content the operator was with his or her execution of the task. The frustration level judges how the operator felt while completing the task.

Figure 6.2 outlines the seven steps to implementing the NASA-TLX. Step 1 of the process is to identify the task subject to analysis. The next step requires breaking down the work elements of the task by conducting a hierarchical task analysis (HTA). During step 3, operators are chosen to meet the needs of the experimental design or process modification to be analyzed. Once the operators have been selected, they should be briefed (step 4) on the aim of the evaluation and the NASA-TLX method. In step 5, the task under analysis is performed by the operator. After the task is completed, the operator starts the NASA-TLX rating procedure. It is recommended that the operators are given the definitions of each MWL dimension with the rating scale to assist with the process. This step includes completing a pairwise comparison and interval scale. The pairwise comparison is a weighting procedure that poses 15 pairwise combinations to the operators and has them choose the scale from each pair that has the most influence on their perceived workload during the task under analysis (Stanton et al., 2010). The most common modification to the NASA-TLX is to exclude the weighting process from the procedure (Hart, 2006). The modified method has been referred to as the *raw NASA-TLX* (RTLX); the subscale ratings are averaged to form an overall workload

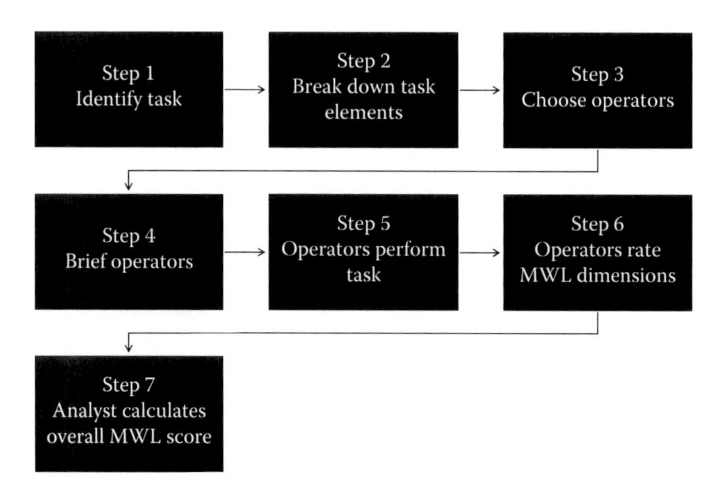

Figure 6.2 NASA-TLX procedural steps.

appraisal. The interval scale rates the six dimensions on a scale of 0–100, where 0 represents the lowest task demand and 100 is highest, with the exception of the performance dimension, where 0 indicates high demand and 100 is low demand. The final action of the process (step 7) is to calculate the overall workload score.

Workload profile

WP is a multidimensional subjective workload measurement tool that is based on the multiple resource model proposed by Wickens (1987) (Stanton et al., 2010; Tsang and Velazquez, 1996; Rubio et al., 2004). This assessment tool utilizes eight dimensions in the collection of data on the demands imposed by a task: perceptual/central processing, response selection execution, spatial processing, verbal processing, visual processing, auditory processing, manual output, and speech output. The WP technique offers an overall rating for the specific task elements as well as a rating for each of the workload dimensions. These multidimensional ratings should provide the analyst with a helpful interpretation of MWL in order to categorize the ways in which the task is demanding (Tsang and Velazquez, 1996).

Figure 6.3 summarizes the nine steps of the WP process according by Stanton et al. (2010). The first step of the process is to specify the tasks or scenario under analysis. Next, an HTA should be administered in order to break down the work elements; this aids in understanding the complete task. Step 3 calls for a workload pro forma to be developed. The pro forma shown in Figure 6.4 lists the task elements to be rated in random order down one column and records the eight MRT workload dimensions across the page. These dimensions are the stages of processing (perceptual/central vs.

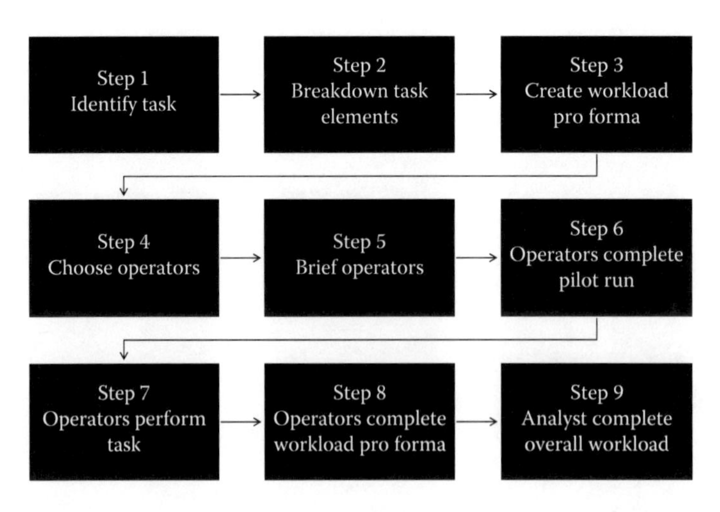

Figure 6.3 WP procedural steps.

Operator #	Workload dimensions							
	Stages of processing		Code of processing		Input		Output	
Work elements	Perceptual	Response	Spatial	Verbal	Visual	Auditory	Manual	Speech
#1								
#2								
#3								
#4								

Figure 6.4 WP pro forma. (Adapted from P. S. Tsang and V. L. Velazquez, *Ergonomics*, *39*(3), 358–381, 1996.)

response), the modalities of input (visual vs. auditory) and output (manual vs. vocal), and the codes of processing (verbal vs. spatial).

The operators are chosen in step 4. In step 5, the operators are briefed on the purpose of the process evaluation. It is recommended that they receive a seminar or lesson on MWL, MWL assessment, MRT, and the WP dimensions, in order to gain a clear understanding of the process, because one disadvantage of the WP method is that understanding the different dimensions is difficult if one has limited experience of psychology and human factors. Once the operators have a clear understanding of the technique's process, in step 6, a pilot run with a small set of work elements should be completed to demonstrate how to complete the profile pro forma. The task under analysis will be performed by the operators in step 7. Step 8 requires the completion of the WP pro forma. During this step, the operators rate the proportion of attentional resources used for each task. Ratings ranging from 0 to 1 for each of the MWL dimensions hypothesized in the Wickens model of MRT are assigned by the operators. During the final step (step 9), the analyst calculates the overall workload for each task by adding the ratings across all of the dimensions.

Multiple Resource Theory The underpinning theory of WP is MRT. MRT is a predictive model that provides an understanding of operator performance abilities while multitasking in a *complex environment* (Wickens, 2002). A complex environment is a task that has a number of concurrent work elements that are time shared (Liu and Wickens, 1988). When an individual performs a task, each operation uses mental processing resources that are vital to carrying out the task (Mitchell, 2003). People have the capacity to multitask until the task demands exceed their available resources. According to MRT, the human mind has various resources that can be allocated to task demands either individually or collectively, including visual, auditory, cognitive, motor, and speech resources. When task demands overlap, available resources are reduced and the MRT model predicts that performance will degrade. When multiple tasks involve competing resources, this competition could lead to a compromise in system safety and effectiveness. So, the principal value of the MRT model is predicting the relative differences in multitasking between various conditions (Wickens, 2008).

The MRT model can predict performance breakdowns in high workload circumstances, such as those that require an operator to multitask by performing two or more work elements at one time. This model can make practical predictions regarding performance breakdowns in a human operator's ability to accomplish tasks—for example, someone driving an automobile in traffic, a pilot landing an aircraft, or a secretary in a busy office (Wickens, 2002). A significant application of MRT is identifying resource overload when multitasking conditions occur. Examples of strategies for reducing resource overload include automating parts of tasks, reassigning task elements to multiple operators, and changing task procedures such that process steps are performed sequentially instead of concurrently (Wickens et al., 2015).

The MRT model uses four dimensions that justify the variances in the time sharing of performance, and each dimension has two discrete levels (Wickens, 2002). Wickens (2008) defines the dimensions as follows:

- Processing stages
- Perceptual modalities
- Visual channels
- Processing codes

During the processing stages, perceptual and cognitive tasks utilize different resources from selective and executive tasks. Individuals are capable of time sharing effectively between activities that are perceptually and cognitively demanding because these resources appear to be the same. The perceptual modalities dimension infers that auditory perception uses different resources from visual perception. The implication of this dimension is that time sharing for cross-modal conditions as opposed to intramodal conditions is superior, because attention can be better distributed between the eye and ear rather than having two visual or two auditory channels, since visual perception uses different resources from auditory perception. Visual channels differentiate between two different types of resources: focal and ambient vision. Focal vision supports object recognition, and ambient vision is responsible for the perception of orientation and the movement of objects. Processing codes indicate spatial activity employs different resources from verbal/linguistic activity. Separating spatial and verbal resources improves efficiency. Understanding the dimensions in the MRT model can aid in evaluating and allocating the various mental resources for different task conditions in work system design. There are simulation tools that are based on MRT and can be used for workload modeling, such as the Improved Performance Research Integration Tool (IMPRINT) and the Man–Machine Integration Design and Analysis System (MIDAS) (Wickens, 2008).

Physiological measures

Physiological workload measures are applied mostly to uninterrupted measurements of the body's physical responses (Miller, 2001). An advantage of these measures is they do not interfere with the primary task performance, but depending on the collection method, there may be a degree of physical constraint imposed (Wickens et al., 2015). The use of physiological measures as an indicator of MWL is based on the assumption that as task demand increases, changes in various participant physical responses are noticeable (Stanton et al., 2010). These physical responses can be measured with various bodily activities, such as

- Cardiac
- Brain
- Respiratory
- Speech
- Eye

Cardiac activity is generally measured with the operator's heart rate, heart rate variability (HRV), and blood pressure. Brain activity can be measured using an electroencephalograph (EEG) and an electro-oculegram (EOG). To measure respiratory activity, the volume of air a person breathes in and the amount of breaths taken in a given time can be used. Speech measures of pitch, rate, and loudness can be taken into account when evaluating MWL. Eye measures can include fixation rate, fixation duration, blink rate, blink duration, pupil diameter, saccade duration, and saccade frequency.

As previously mentioned, the use of physiological measures can be evaluated by various methods; however, this review will focus on eye activity. For over 20 years, psychologists have argued that pupillary changes are indications of cognitive load, and many studies have validated this argument (Marshall, 2002). Fixations are moments when the eyes are spent looking at an object. Blink rate is the number of times the eyes close in a specified time. Blink duration is an interval of eye closure defined as the length of time spent blinking. Pupil diameter has been found to increase with workload: the size of the pupil increases as a higher workload is experienced. Saccade is involuntary eye movement between fixations. According to Ahlstrom and Friedman-Berg (2006), research findings indicate that blink rate and blink duration decline as workload decreases, pupil dilation increases as cognitive demand increases, and an increase in the number of saccades or a decrease in saccade duration are relative to increased workload. Also, increased fixation duration can be interpreted as an indicator of increased MWL. Doherty et al.'s (2010) findings show that the frequency of long fixations increases with increased workload.

In order to collect the eye measurements, the process of eye tracking is utilized. Eye tracking is a technology used for measuring eye activity: measuring where an individual is looking (i.e., gaze direction) at any given time, the sequence in which the eyes are shifting from one location to another, or how much the pupil dilates while looking. An eye tracker is a device that deals with the measuring of eye positions and eye movement. Eye trackers are used to evaluate human–machine interaction. The eye-tracking data are collected using either a remote, head-mounted, or wearable eyeglass system. These data can be utilized to collect information relative to numerous gaze measurements. Since eye movements are commonly considered to be involuntary, eye tracking provides objective data of users' visual interaction with a system (Bruneau et al., 2002).

Performance

MWL can be measured by monitoring an operator's performance and observing how it varies with changes in task demands (Hertzum and Holmegaard, 2013). Estimations of MWL can be determined by measuring a person's performance on a task and evaluating whether his or her performance degrades with increasing workload (Miller, 2001). When considering performance in repetitive task processes such as manufacturing operations, not only should the calculated measures of the operator's performance be utilized to evaluate MWL, but unconscious cognition and learning curves (LCs) should also be factored into the system's design.

Calculated measures

The key categories of performance-based workload measures are primary and secondary task performance. Primary task performance measures the operator's adequacy in the main task, while secondary task performance evaluates the workload of the primary task of interest by assessing the operator's ability to perform a second task in conjunction with the primary task. According to Miller (2001), primary task performance should be used to measure MWL, because it is easier to measure than secondary task performance and it has been more extensively studied for accuracy. Also, it is expected that the primary task performance of the operator will reduce with increased MWL for the task under analysis (Stanton et al., 2010).

The MWL necessitated for the completion of a given task can be assessed with performance metrics such as time, speed, or number of errors (Cegarra and Chevalier, 2008). As cited by Cegarra and Chevalier (2008), studies have shown a correlation between poor performance in the main task and the MWL experienced by participants. Bubb (2005) defined human error probability (HEP) as the likelihood that a task would be defectively completed during a specified timeframe. HEP is an accuracy measurement expressed mathematically as follows:

$$\text{HEP} = \frac{\text{number of observed errors}}{\text{number of the possiblities for an error}}$$

Galy et al. (2012) cite a study by Fournier et al. (1999) in which the authors examined subjects in a multitasking experiment, and the results revealed that performance decreased as workload increased. This implies that a decline in performance can be an indicator of MWL overload. Therefore, the HEP metric should increase as MWL increases. The HEP metric is only one example of a performance measure. Others can be defined depending on the goal of the process design.

Unconscious cognition

In repetitive task operations such as production processes, there is a process of learning. Learning is a natural phenomenon in which a worker's output improves as his or her productivity levels improve (Badiru, 1995). As an operator becomes more efficient in a process over time and the learning reduces, an operator can experience *unconscious cognition*. Unconscious cognition is the "mental structure and process that, operating outside phenomenal awareness, nevertheless influences conscious experiences, thought and action" (Kihlstrom, 1987). The actions required to drive an automobile are an example of unconscious cognition. These actions, which are completed regularly without apparent mental effort or recollection, are considered by some to be done unconsciously (Greenwald, 1992). If one does experience unconscious cognition, according to Reinhard et al. (2013), unconscious processes are presumed to have far better processing capabilities than conscious thinking, and as a result of this unconscious state, one should outperform conscious thought when decision-making problems are complex. Therefore, being able to design a work system to achieve unconscious cognition for the operator would be most desirable.

Learning

"Learning refers to the improved operational efficiency and cost reduction obtained from repetition of a task" (Badiru, 2012a). Workers learn and improve work performance by repeating tasks over time. As workers become more skilled and develop higher capabilities, a larger mental resource supply is effectively created, improving the efficiency of the operator (Vidulich and Tsang, 2012). Therefore, their output levels should improve with learning, which decreases expenditures per unit and, believably, improves product quality (Badiru, 1995). However, a worker's performance can also regress as a result of forgetting and work interruptions. Forgetting is an outcome of work interruptions, which is made apparent by production rate reductions and the manufacturability of lower-quality products after an inoperative period compared with production during continuous operation (Jaber and Bonney, 2011). Therefore, the ability to predict an operator's performance when work resumes is valuable because this can improve production planning and resource allocation. In addition, incorporating learning into a production process decreases fatigue, which can also improve the performance of the system (Jaber et al., 2013).

"The learning curve phenomenon has been of interest to researchers and practitioners for many years" (Jaber and Bonney, 2011). LCs are used to determine how fast a skill can be mastered. However, industrial LCs can be used for production planning and control, and there is a large body of literature examining in how processes have been improved as a result of LCs. LCs have been used in manufacturing for decision making, manpower scheduling, resource allocation, cost optimization, and more. Although learning is a well-examined topic, this concept should be further extended to studies and experiments in order to understand how forgetting affects MWL. From the literature, it is known that after a break, the first cycle of work is the longest; and, according to MWL theory, performance (e.g., time on task) is an indication of cognitive load. This correlates to an operator experiencing the greatest load at the first cycle of work. Using LCs to determine learning profiles could assist in reducing an operator's cognitive load at production start-up.

In order to model a worker's output, LCs can be utilized. LC analysis can be useful for predicting expected work rates in processes where output accountability is tracked (Badiru, 2014). In jobs with repetitive tasks, LCs can mathematically depict a worker's performance (Jaber and Bonney, 2011). According to Jaber and Bonney (2011), this method is deemed to be an effective tool to monitor workers' performances in repetitive tasks, leading to a reduction in process loss due to their inability in the first production cycles. LCs can be applied to the allocation of tasks so as to be consistent with workers' learning profiles, in order to control productive operations. Other applications of LCs include production planning, cost estimation and control, resource allocation, lot sizing, and product pricing (Badiru, 1995). High performance with low mental effort is the most efficient combination of cognitive capacity, which can result from the theory of learning.

Summary

Processes should be evaluated to confirm whether high levels of MWL are within workers' capabilities, such that the average workload is not so high that it creates cumulative mental fatigue and not so low that operators are bored and may miss important system cues (Badiru and Resnick, 2013). There are various MWL measures that can be used to confirm workload levels imposed by the system. These MWL measures are often used to verify whether the human is operating within a tolerable information-processing capacity while executing task functions and elements. As technology continues to change, there is a constant need to know its impact on the human operator, in order to design safe and highly productive operations for work performance. Workload is the quantity of resources applied in the performance of a specific task (Boff and Lincoln, 1988). The mental resources of humans are limited; therefore, it is important that the work performed does not overload human capacity, which could lead to performance degradation. In order to optimize MWL in efforts to improve performance, the analyst must have accurate predictions of the performance. There are three categories of MWL measures: subjective, physiological, and performance. No single measure of MWL evaluates all of the dimensions of mental load; so, when possible, use a combination of measures from the three major categories. However, for maximum accuracy, it is recommended that one measure be continuous, one measure be taken at intervals during the experiment, and one measure be taken after the experiment (Miller, 2001). Once an accurate MWL prediction is obtained for a task, these predictions can be used for the overall capacity planning of work systems by identifying the need for MWL reductions, as well as determining whether additional work elements can be assigned to operators without degrading performance. MRT can be employed to determine which mental resources were used in an effort to enhance the process design by restructuring mental resource utilization in order to improve operator performance.

References

Ahlstrom, U., and Friedman-Berg, F. J. (2006). Using eye movement activity as a correlate of cognitive workload. *International Journal of Industrial Ergonomics*, *36*(7), 623–636.

Badiru, A. (2014). Adverse impacts of furlough programs on employee work rate and organizational productivity. *Defence ARJ*, *21*(2), 595–624.
Badiru, A. B. (1995). Multivariate analysis of the effect of learning and forgetting on product quality. *The International Journal of Production Research*, *33*(3), 777–794.
Badiru, A. B. (2012a). Half-life learning curves in the defense acquisition life cycle. *Defense Acquisition Research Journal*, 19(3), 283–308.
Badiru, A. B. (2012b). Half-life of learning curves for information technology project management. In J. Wang (ed.), *Project Management Techniques and Innovations in Information Technology*, pp. 146–164. Hershey, PA: IGI Global.
Badiru, A. B., and Resnick, M. (2013). Human factors. In A. B. Badiru (ed.), *Handbook of Industrial and Systems Engineering* (2nd ed.), pp. 431–454. Boca Raton, FL: CRC Press.
Bertram, D. A., Opila, D. A., Brown, J. L., Gallagher, S. J., Schifeling, R. W., Snow, I. S., et al. (1992). Measuring physician mental workload: Reliability and validity assessment of a brief instrument. *Medical Care*, *30*(2), 95–104.
Boff, K. R., and Lincoln, J. E. (1988). *Engineering Data Compendium: Human Perception and Performance*, Vol. 3. Wright-Patterson A.F.B., OH: Harry G. Armstrong Aerospace Medical Research Laboratory.
Bommer, S. C., and Fendley, M. (2015). Assessing the effects of multimodal communications on mental workload during the supervision of multiple unmanned aerial vehicles. *International Journal of Unmanned Systems Engineering*, *3*(1), 38.
Bruneau, D., Sasse, M. A., and McCarthy, J. (2002). The eyes never lie: The use of eye tracking data in HCI research. *Proceedings of the CHI*, *2*, 25.
Bubb, H. (2005). Human reliability: A key to improved quality in manufacturing. *Human Factors and Ergonomics in Manufacturing & Service Industries*, *15*(4), 353–368.
Cao, A., Chintamani, K. K., Pandya, A. K., and Ellis, R. D. (2009). NASA TLX: Software for assessing subjective mental workload. *Behavior Research Methods*, *41*(1), 113–117.
Cegarra, J., and Chevalier, A. (2008). The use of tholos software for combining measures of mental workload: Toward theoretical and methodological improvements. *Behavior Research Methods*, *40*(4), 988–1000.
Colombi, J. M., Miller, M. E., Schneider, M., McGrogan, M. J., Long, C. D. S., and Plaga, J. (2012). Predictive mental workload modeling for semiautonomous system design: Implications for systems of systems. *Systems Engineering*, *15*(4), 448–460.
Cooper, G. E., and Harper Jr, R. P., (1969). The use of pilot rating in the evaluation of aircraft handling qualities. No. AGARD-567. Advisory Group for aerospace research and development. Neuilly-Sur-Seine, France.
DiDomenico, A., and Nussbaum, M. A. (2008). Interactive effects of physical and mental workload on subjective workload assessment. *International Journal of Industrial Ergonomics*, *38*(11), 977–983.
DiDomenico, A., and Nussbaum, M. A. (2011). Effects of different physical workload parameters on mental workload and performance. *International Journal of Industrial Ergonomics*, *41*(3), 255–260.
Doherty, S., O'Brien, S., and Carl, M. (2010). Eye tracking as an MT evaluation technique. *Machine Translation*, *24*(1), 1–13.

Endsley, M. R., and Rodgers, M. D. (1997). *Distribution of Attention, Situation Awareness, and Workload in a Passive Air Traffic Control Task: Implications for Operational Errors and Automation*. Washington, DC: US Department of Transportation, Federal Aviation Administration, Office of Aviation Medicine.

Estes, S. (2015). The workload curve: Subjective mental workload. *Human Factors*, *57*(7), 1174–1187.

Fournier, L. R., Wilson, G. F., and Swain, C. R. (1999). Electrophysiological, behavioral, and subjective indexes of workload when performing multiple tasks: Manipulations of task difficulty and training. *International Journal of Psychophysiology*, *31*(2), 129–145.

Galy, E., Cariou, M., and Mélan, C. (2012). What is the relationship between mental workload factors and cognitive load types? *International Journal of Psychophysiology*, *83*(3), 269–275.

Greenwald, A. G. (1992). New look 3: Unconscious cognition reclaimed. *American Psychologist*, *47*(6), 766.

Hart, S. G. (2006). NASA task load index (NASA-TLX): 20 years later. *Proceedings of the Human Factors and Ergonomics Society Annual Meeting*, *50*(9), 904–908.

Hart, S. G., and Staveland, L. E. (1988). Development of NASA-TLX (task load index): Results of empirical and theoretical research. *Advances in Psychology*, *52*, 139–183.

Hertzum, M., and Holmegaard, K. D. (2013). Perceived time as a measure of mental workload: Effects of time constraints and task success. *International Journal of Human–Computer Interaction*, *29*(1), 26–39.

Jaber, M., Givi, Z., and Neumann, W. (2013). Incorporating human fatigue and recovery into the learning: Forgetting process. *Applied Mathematical Modelling*, *37*(12), 7287–7299.

Jaber, M. Y., and Bonney, M. (2011). The lot sizing problem and the learning curve. In M. Y. Jaber (ed.), *Learning Curves: Theory, Models, and Applications*, pp. 265–291. Boca Raton, FL: CRC Press.

Jung, H. S., and Jung, H. (2001). Establishment of overall workload assessment technique for various tasks and workplaces. *International Journal of Industrial Ergonomics*, *28*(6), 341–353.

Kihlstrom, J. F. (1987). The cognitive unconscious. *Science*, *237*(4821), 1445–1452.

Lindblom, J., and Thorvald, P. (2014). Towards a framework for reducing cognitive load in manufacturing personnel. *Advances in Cognitive Engineering and Neuroergonomics*, *11*, 233–244.

Liu, Y., and Wickens, C. (1988). Patterns of task interference when human functions as a controller or a monitor. *Proceedings of the IEEE International Conference on Systems, Man, and Cybernetics*, *1988*, 2, 864–867.

Loft, S., Sanderson, P., Neal, A., and Mooij, M. (2007). Modeling and predicting mental workload in en route air traffic control: Critical review and broader implications. *Human Factors*, *49*(3), 376–399.

Marshall, S. P. (2002). The index of cognitive activity: Measuring cognitive workload. *Proceedings of the 2002 IEEE 7th Conference on Human Factors and Power Plants*, *2002*, 75–79.

May, J. G., Kennedy, R. S., Williams, M. C., Dunlap, W. P., and Brannan, J. R. (1990). Eye movement indices of mental workload. *Acta Psychologica*, *75*(1), 75–89.

Miller, S. (2001). *Workload Measures*. Iowa City, IA: National Advanced Driving Simulator.

Mitchell, D. K. (2003). *Advanced Improved Performance Research Integration Tool (IMPRINT) Vetronics Technology Test Bed Model Development*. Aberdeen Proving Ground, MD: US Army Research Laboratory.
Murata, A. (2000). Ergonomics and cognitive engineering for robot–human cooperation. *Proceedings of the 9th IEEE International Workshop on Robot and Human Interactive Communication (RO-MAN), 2000*, 206–211.
Pesante, J. A., Williges, R. C., and Woldstad, J. C. (2001). The effects of multitasking on quality inspection in advanced manufacturing systems. *Human Factors and Ergonomics in Manufacturing & Service Industries*, *11*(4), 287–298.
Reid, G. B., and Nygren, T. E. (1988). The subjective workload assessment technique: A scaling procedure for measuring mental workload. *Advances in Psychology*, *52*, 185–218.
Reinhard, M., Greifeneder, R., and Scharmach, M. (2013). Unconscious processes improve lie detection. *Journal of Personality and Social Psychology*, *105*(5), 721.
Roscoe, A. H., and Ellis, G. A. (1990). *A Subjective Rating Scale for Assessing Pilot Workload in Flight: A Decade of Practical Use*. Bedford, UK: Royal Aerospace Establishment.
Rubio, S., Díaz, E., Martín, J., and Puente, J. M. (2004). Evaluation of subjective mental workload: A comparison of SWAT, NASA-TLX, and workload profile methods. *Applied Psychology*, *53*(1), 61–86.
Stanton, N. A., Hedge, A., Brookhuis, K., Salas, E., and Hendrick, H. W. (2004). *Handbook of Human Factors and Ergonomics Methods*. Boca Raton, FL: CRC Press.
Stanton, N. A., Salmon, P. M., Walker, G. H., Baber, C., and Jenkins, D. P. (2010). *Human Factors Methods: A Practical Guide for Engineering and Design* (1st ed.). Burlington, VT: Ashgate.
Tsang, P. S., and Vidulich, M. A. (2006). Mental workload and situation awareness. *Handbook of Human Factors and Ergonomics* (3rd ed.), pp. 243–268. Hoboken, NJ: Wiley.
Vidulich, M. A., and Tsang, P. S. (2012). Mental workload and situation awareness. In G. Salvendy (ed.), *Handbook of Human Factors and Ergonomics*, pp. 243–273. Hoboken, NJ: Wiley.
Vidullch, M. A., Ward, G. F., and Schueren, J. (1991). Using the subjective workload dominance (SWORD) technique for projective workload assessment. *Human Factors: The Journal of the Human Factors and Ergonomics Society*, *33*(6), 677–691.
Wickens, C., and Carswell, C. (2006). *Handbook of Human Factors and Ergonomics*. Hoboken, NJ: Wiley.
Wickens, C. D., and Hollands, J. G. (2000). Attention, time-sharing, and workload. *Engineering Psychology and Human Performance* (3rd ed.), pp. 439–479. Upper Saddle River, NJ: Prentice Hall.
Wickens, C. D. (2002). Multiple resources and performance prediction. *Theoretical Issues in Ergonomics Science*, *3*(2), 159–177.
Wickens, C. D. (2008). Multiple resources and mental workload. *Human Factors*, *50*(3), 449–455.
Wickens, C. D., Hollands, J. G., Banbury, S., and Parasuraman, R. (2015). *Engineering Psychology & Human Performance*. New York: Psychology Press.
Xie, B., and Salvendy, G. (2000). Prediction of mental workload in single and multiple tasks environments. *International Journal of Cognitive Ergonomics*, *4*(3), 213–242.

section four

Work justification

chapter seven

Cognitive modeling for human performance predictions

This chapter introduces the cognitive modeling of human performance in process design and modification as a method to justify the mental load of work design. Modeling human cognition and understanding the means by which humans handle information is becoming more and more significant as system designers develop automation to aid human operators with tasks that were traditionally manually controlled and physical in nature and which are being substituted with tasks that are cognitive in nature (Gore et al., 2008). As work systems are becoming gradually more automated, the relevance of designing a system with humans in mind is crucial. From air traffic control to automobiles, how adequately operators perform relative to what the system expects of them will impact how well the system ultimately functions (Laughery et al., 2000). There are two key elements to a safety-critical system: the human and the machine (Greoriades and Sutcliffe, 2008). The interaction between humans and other system elements, such as hardware, software, tasks, environments, and work structures, is a human–machine system. This system could be simple, such as an operator working with a hand tool, or it may be complex, such as an aviation system. There are three factors that typically have a major effect on system performance in any human–machine system: human error, system design, and human limitations (Boff and Lincoln, 1988). A method for predicting the effects of these factors on the human–machine system is human performance modeling (HPM), because modeling tools apply well-established theoretical concepts of human information processing to produce performance predictions (Kieras and Meyer, 2000). Some HPM tools that incorporate cognitive modeling for work system design and modification are covered in this chapter.

Human performance modeling

The term HPM is used to define systems that simulate human decision making and actions, which are basically synonymous with cognitive simulation and artificial intelligence (Boring et al., 2015). HPM is a computer-aided work analysis software methodology employed to predict complex human–automation integration and system flow patterns, with the aim of increasing operator and system safety (Gore, 2002). According to Nan and Sansavini (2016), the first human performance–related model was developed in the early 1980s, and since that time many models and approaches have been developed to assess human performance.

The process of HPM includes inserting human attributes within a computer software construct that simulates a human operator interacting with an operating environment (Gore and Milgram, 2013). It can be utilized to illustrate and predict human behaviors working under various conditions (Jiang et al., 2013). One of the primary advantages of modeling is that it lets analysts assess different conditions by altering various system parameters, such as the working conditions, equipment design, and manpower allocation, to determine the effects of these parameters on the predicted performance; such analyses can be applied to establish the requirements for equipment designs, staffing configurations, training, task designs, or scheduling (Sebok et al., 2015). On the other hand, in order to evaluate total performance, parameters such as the number of converging measures, task latency, the type and probability of errors, the quality of performance, and workload measures should be evaluated (Laughery et al., 2012).

According to Laughery et al. (2012), human system modeling can be utilized to address many questions as a function of the system design, task allocation, individual capabilities, and environment. Figure 7.1 outlines some of those questions as described by Laughery et al. (2012).

Methods for analyzing human performance in the design stage include prototyping, human-in-the-loop (HITL) simulation, and human-out-of-the-loop (HOOTL) simulation. Prototyping is the process of developing a pilot form (i.e., a prototype) of a system in order to refine the requirements of the system or to uncover critical design elements (Gordon and Bieman, 1995). Prototyping can give both the engineer and the user an opportunity to evaluate the system in order to assess whether it meets the user's needs. Engineers can also use this method to improve their knowledge of the technical demands and feasibility of a proposed system. An HITL simulation is a modeling framework that involves human interaction (Rothrock and Narayanan, 2011). HITL simulations give researchers and analysts the ability to study the complexities of a system with human involvement from a holistic perspective. The disadvantages of these methods (prototyping and HITL) are the time associated with experimentation and the cost of building functioning prototypes for an iterative process. HOOTL simulation is a methodology that makes use of computer models of human performance by creating a virtual human agent that works collectively with new technologies and procedures (Gore, 2002). HOOTL simulations can be utilized in the initial stage of the development process of a product, system, or technology to create procedures and training requirements. In addition, HOOTL simulations can be applied in the

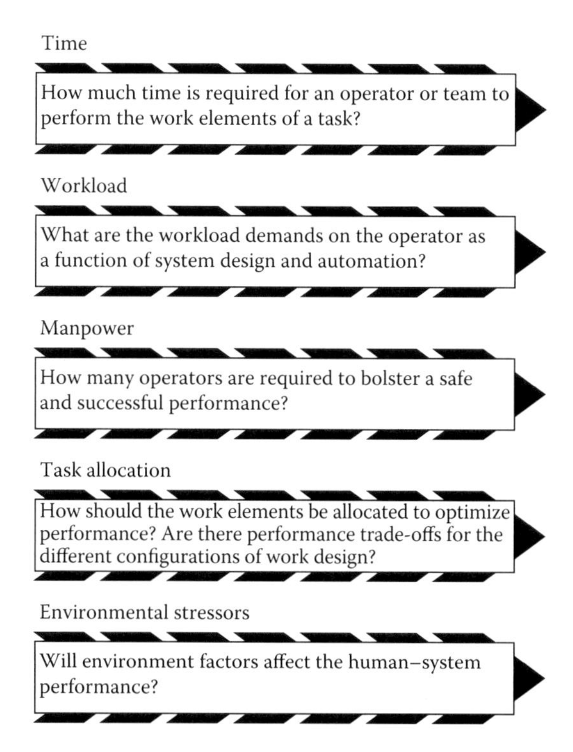

Time
How much time is required for an operator or team to perform the work elements of a task?
Workload
What are the workload demands on the operator as a function of system design and automation?
Manpower
How many operators are required to bolster a safe and successful performance?
Task allocation
How should the work elements be allocated to optimize performance? Are there performance trade-offs for the different configurations of work design?
Environmental stressors
Will environment factors affect the human–system performance?

Figure 7.1 Questions that can be answered with human system modeling.

assessment of operator safety, operator productivity, and efficient system design to identify vulnerabilities where potential human–system errors are likely to occur. The use of HOOTL simulations is more cost and time efficient than waiting for the concept to be fully designed and used in practice, as with HITL tests. One criticism of HOOTL methods is that the simulation software only predicts input–output behavior in mechanistic terms (Gore, 2002). Therefore, it is imperative to understand the system structure and define the appropriate inputs for the system.

HPM can be used synergistically with HITL studies, as shown in Figure 7.2, in the following circumstances (Gore and Milgram, 2013):

1. The development of complex systems that include human operators interacting with existent technology and automation to complete numerous interacting and conflicting tasks with time constraints
2. Times when events cannot be investigated fully with HITL subjects as a result of safety concerns
3. Projects limited by cost constraints
4. Difficult situations associated with simulating unusual events

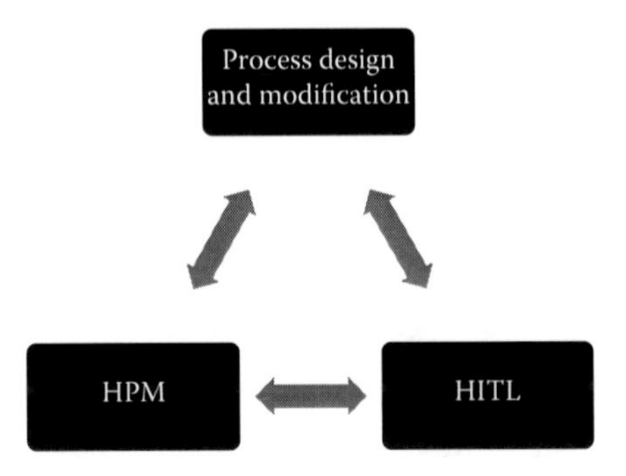

Figure 7.2 Modeling synergy in process design and modification.

An example of applying HPM and HITL synergy is Wickens et al. (2015). This study used both HPM and HITL to validate model predictions relative to differences in complacency across various stages of automation by simulating operators working on a robotic arm task in space missions.

Figure 7.2 illustrates how HPM and HITL methods can support both process design and modification collectively. HITL studies can be used to validate human performance models. The information obtained from HITL studies can in turn be utilized to modify the model design for better performance predictions. On the other hand, HPM can be used to design better experiments for process design validation. The use of models for predicting human performance is an efficient approach to process design and modification. Also, accurate human performance predictions can be valuable in implementing staff recruitment and personnel strategies in an effort to improve competitiveness in manufacturing (Li et al., 2016).

Modeling systems

HPM systems conceptualize work design. A few of these are as follows:

- Adaptive Control of Thought—Rational (ACT-R; Anderson et al., 1996)
- Man–Machine Integration Design and Analysis System (MIDAS; Staveland, 1994)
- Distributed Operator Model Architecture (D-OMAR; Deutsch et al., 1995)
- Attention–Situation Awareness (A-SA; Wickens et al., 2008)

The following section provides a brief description of seven HPM systems that integrate cognitive modeling elements.

1. *Micro Saint* is a commercial product that supports the development of task network models (Laughery et al., 2000). Task network modeling is an approach to modeling human performance that organizes task sequences in a manner to represent the human–machine system. In a task network model, the performance of a function by a human operator is disintegrated into a series of subfunctions, which are then broken down into tasks (Laugherty et al., 2000). Task network modeling can assist with examining designs for various types of issues, such as how long it takes to perform a task and the potential error rate.
2. *State, Operator, and Result (SOAR)* is a general intelligence architecture developed by Laird et al. (1987). It is based on chunking theory, which is a learning scheme that integrates learning and performance. This method incorporates existing problem-solving knowledge with knowledge of results in a given problem space and transforms it into new knowledge to be used in future problem solving (Laird et al., 1987). The output from SOAR is a comprehensive simulation from which a range of performance measures can be obtained (Pew, 2008).
3. *Goals—Operator, Methods, and Selection Rules (GOMS)* is a systematic description of how to calculate the time needed to perform a task by accounting for the physical and mental work essential to completing the task (Pew, 2008). It identifies the operator's goals, breaks down these goals into subgoals, and demonstrates how they are achieved through operator interaction to give a description of human performance (Baber et al., 2013). The goals are what the user plans to achieve; operators are actions executed while working toward a goal; methods are the sequences of operators and subgoals invoked to accomplish a goal; and selection rules represent the user's knowledge of which method to exercise when there are multiple ways to accomplish the goal (Oyewole and Haight, 2011). This method was originally developed by Card et al. (1983).
4. *Executive-Process Interactive Control (EPIC)* is a computer simulation modeling tool that was developed by Kieras and Meyer (1995) for modeling human cognition and performance. EPIC can provide an in-depth representation of the perceptual, motor, and cognitive constraints on a human operator's ability to perform tasks (Kieras and Meyer, 1997). It was originally developed to evaluate human performance limitations on the effects of an interface design that supports complex task performance in multimodal time-stressed domains. This cognitive architecture is most suited to modeling human multimodal and multitasking performance (Meyer et al., 2001; Kieras and Meyer, 1997).
5. *ACT-R*, as developed by Anderson (1993), is a model of higher-level cognition (Anderson et al., 1997). It is a cognitive architecture that

integrates theories of cognition, visual attention, and motor movement. ACT-R has been used successfully to model higher-level cognition developments such as working memory, scientific reasoning, and skill acquisition (Ritter and Kim, 2015). It has been applied to build models in domains such as human–computer interaction, to generate user models that can evaluate different computer interfaces; education (cognitive tutoring systems), to predict the problems that students may have and offer focused help; and computer-generated forces, to offer cognitive agents that inhabit training environments; to name a few (Budiu, 2013). One vital aspect of ACT-R that differentiates it from other similar theories is that it permits researchers to gather quantitative measures that can be directly evaluated against the quantitative measures collected from human participants (Budiu, 2013).

6. *MIDAS* is an HPM tool for conducting human factor analyses of complex human–machine systems that was developed at the NASA Ames Research Center to evaluate procedures, controls, and displays prior to applying more expensive and time-consuming hardware simulators and human subject experiments (Tyler et al., 1998). The development of MIDAS was a joint effort between the US Army and NASA called the A^3I Program, which was initiated in 1985. The aim of the A^3I Program was to support the exploration of computational representations of human–machine performance to aid designers of crew systems (Corker and Smith, 1993). The MIDAS Task Loading Model (MIDAS-TLM) was originally designed to computationally model the demands that various designs impose on operators to aid engineers in the conceptual design of aircraft crew stations (Staveland, 1994). The overall goal of the MIDAS-TLM was to predict the information-processing demands on the individual four psychological dimensions (i.e., visual, auditory, cognitive, and motor) of the system's operator associated with the tasks of conceptual system design (Staveland, 1994). The demands are used to calculate an estimate of task loading.
7. *Improved Performance Research Integration Tool (IMPRINT)* is HPM software that is used as a discrete event simulation tool to predict mental workload (MWL). Its underpinning theory is based on multiple resource theory (MRT), as discussed in Chapter 6. IMPRINT has analytical capabilities including "human versus system function allocation, mission effectiveness modeling, maintenance manpower determination, mental workload estimation, prediction of human performance under extreme conditions and assessment of performance as a function of varying personnel skills and abilities" (Allender, 2000). This HPM method was developed by the Human Research and Engineering Directorate (HRED) of the US Army

Research Laboratory (ARL) in the 1980s (Mitchell, 2000). IMPRINT's primary domain usage has been in the military (Mitchell, 2003; Krausman et al., 2005; Hunn and Heuckeroth, 2006; Chen and Terrence, 2009). IMPRINT models various scenarios, allowing human capabilities to be analyzed relative to MWL predictions, in order to develop a reliable human–system design that uses the optimal range of mental resources. IMPRINT does not provide recommendations on how to mitigate mental overload to enhance the system. However, based on the IMPRINT predictions, the mental resource demands of these different scenarios can be mapped to the associated MRT ratings in IMPRINT for the development of mitigation strategies in work design. Discrete events, with their estimated task times, and MRT ratings, for their associated mental resources, are simulation inputs. IMPRINT has a MWL component that classifies workload relative to the MRT dimensions (Wickens, 2008). With this information, IMPRINT formulates MWL predictions for various task–design combinations.

The Directory of Design Support Methods provides a broad list of computer-based modeling techniques for designing a system or appraising system design. Table 7.1 provides a reference list of applications in various domains, with supporting HPM methods.

Hierarchical task analysis

In order to properly model a man–machine system, a critical step is to identify the key elements of the system. Hierarchical task analysis (HTA) is a technique that can be utilized to determine the process elements as inputs to the model. It was first developed in the 1960s for training process control tasks in the steel and petrochemical industries (Annett, 2003). HTA is a step-by-step process used to break down a task or system work elements, so that the task or system can be examined to any specified level of detail. Actual or potential performance errors can then be identified; however, it does not provide diagnostic tools for these problems.

HTA has seven primary key steps for implementation, as shown Figure 7.3. The first step of the process is to decide on the purpose of the analysis. This is important because it helps define the depth of information to collect and the collection procedures. The next step (step 2) is to solicit all relevant stakeholders to establish an agreement on defining the task goals. This is vital to the analysis for establishing performance goals and knowing when those goals are met. Step 3 identifies sources of information and data acquisition. Preferred sources of data will vary depending on the analysis. Interviews, direct observations, and performance data are examples of data sources. In step 4, a draft of the work system decomposition

Table 7.1 Cognitive modeling studies in HPM

Author(s)	Year	Method		Domain
Bommer	2016	IMPRINT	MWL modeling	Manufacturing
Gore and Smith	2006	MIDAS	Test predictions of simulated operator workload	International Space Station (ISS)
Kieras et al.	2016	EPIC	Cognitive architecture suited for modeling human performance of multimodal and multitasking work systems	Communications (i.e., two-talker speech perception task)
Nan and Sansavini	2016	Cognitive Reliability Error Analysis Method (CREAM)	A human reliability analysis method that integrates cognitive models to assess human performance	Infrastructure systems (e.g., power supplies, telecommunications, rail transportation systems)
Oyewol and Haigt	2011	GOMS	Cognitive modeling for human–computer interaction	Website design
Paik and Pirolli	2015	ACT-R	Cognitive architecture used to simulate the cognitive, perceptual, and motor behavior of humans, including strategy selection	Simulated geospatial (map-based) intelligence tasks
Zhong et al.	2012	SOAR	Cognitive process models for problem solving	Traffic tasks (i.e., traffic guidance compliance behavior)

diagram or table is developed. A key feature of HTA is to compare what the operator is actually doing with why he or she does it and to identify the drawbacks if the task is not performed correctly. When this is understood, a meaningful table or diagram can be created to illustrate the functional structure of the system. Step 5 is the process of cross-checking the work decomposition diagram or table with the stakeholders to ensure its reliability for system modeling. This could be an iterative process. During step 6, important operations are identified regardless of the purpose of the analysis. As a general rule of thumb, the number of task levels in the analysis should cease when the probability of system performance failure

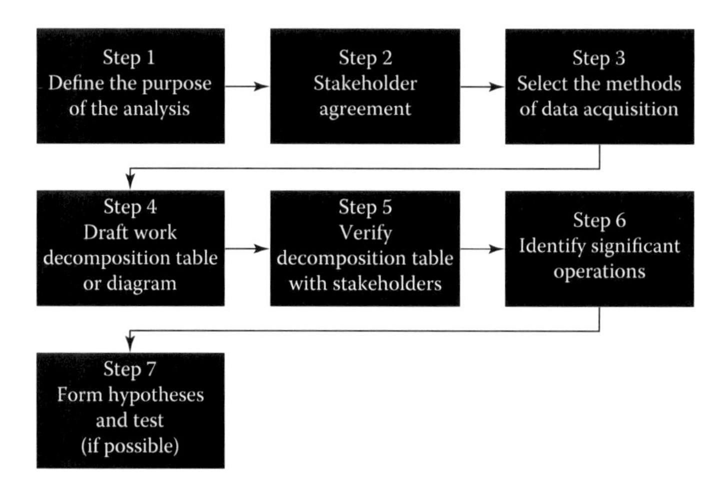

Figure 7.3 Procedural steps for implementing HTA. (Adapted from J. Annett, *Handbook of Cognitive Task Design*, 2, 17–35, 2003.)

multiplied by the cost of failure is acceptably low (Annett, 2003). The final step (step 7) for the analyst is to develop hypotheses for proposing practical solutions in order to account for the failures within. However, these hypotheses must be tested to prove their validity.

In summary, HTA works by disintegrating the system into a hierarchy of goals, subgoals, operations, and plans; it emphasizes what the operator is expected to do in terms of actions and/or cognitive processes to attain a system goal (Salmon et al., 2010). It is a simplistic process that encompasses reviews of the task or system through observation, surveys, interviews, walkthroughs, and user trials, to name a few. The information gathered from these studies is used to break down and describe the goals and subgoals involved in the work system. The process of developing HTA helps the analyst to understand the task or system. The output of an HTA is tremendously useful in identifying the inputs for numerous human factor analyses, such as error analysis, interface design and evaluation, and the allocation of function analysis (Stanton et al., 2010).

Summary

With the aim of improving efficiency, HPM is a valuable method of supporting work system design. It can assist with understanding the interaction between the human and the other system elements. HPM can provide very early predictions of performance in the development phase, in comparison with prototyping and HITL testing. There are numerous HPM methods to choose from, and this chapter has attempted to highlight a few methods that are usable for cognitive modeling in work design.

References

Allender, L. (2000). Modeling human performance: Impacting system design, performance, and cost. *Simulation Series*, 32(3), 139–144.

Anderson, J. R. (1993). *Rules of the Mind*. Hillsdale, NJ: Lawrence Erlbaum Associates.

Anderson, J. R., Matessa, M., and Lebiere, C. (1997). ACT-R: A theory of higher level cognition and its relation to visual attention. *Human–Computer Interaction*, 12(4), 439–462.

Anderson, J. R., Reder, L. M., and Lebiere, C. (1996). Working memory: Activation limitations on retrieval. *Cognitive Psychology*, 30(3), 221–256.

Annett, J. (2003). Hierarchical task analysis. *Handbook of Cognitive Task Design*, 2, 17–35.

Baber, C., Jenkins, D. P., Walker, G. H., Rafferty, L. A., Salmon, P. M., and Stanton, N. A. (2013). *Human Factors Methods: A Practical Guide for Engineering and Design*. Burlington, VT: Ashgate.

Boff, K. R., and Lincoln, J. E. (1988). *Engineering Data Compendium: Human Perception and Performance*, Vol. 3. Wright-Patterson Air Force Base, OH: Harry G. Armstrong Medical Research Lab.

Bommer, S. (2016). A theoretical framework for evaluating mental workload resources in human systems design for manufacturing operations. https://etd.ohiolink.edu/ (accessed February 2017).

Boring, R. L., Joe, J. C., and Mandelli, D. (2015). Human performance modeling for dynamic human reliability analysis. *Proceedings of the International Conference on Digital Human Modeling and Applications in Health, Safety, Ergonomics and Risk Management*, pp. 223–234.

Budiu, R. (2013). ACT-R. Carnegie Mellon University. http://act-r.psy.cmu.edu (accessed September 17, 2016).

Card, S. K., Newell, A., and Moran, T. P. (1983). *The Psychology of Human–Computer Interaction*. Hillsdale, NJ: L. Erlbaum.

Chen, J., and Terrence, P. (2009). Effects of imperfect automation and individual differences on concurrent performance of military and robotics tasks in a simulated multitasking environment. *Ergonomics*, 52(8), 907–920.

Corker, K. M., and Smith, B. R. (1993). An architecture and model for cognitive engineering simulation analysis: Application to advanced aviation automation. *Proceedings of the AIAA Computing in Aerospace 9 Conference*, pp. 1079–1088.

Dean, T. F. (1998). Directory of Design Support Methods, Technical Report ADA355192. San Diego, CA: Defense Technical Information Center, Matris Office.

Deutsch, S. E., Cramer, N. L., and Feehrer, C. E. (1995). *Research, Development, Training, and Evaluation (RDT&E) Support: Operability Model Architecture*. Cambridge, MA: Bolt, Beranek, and Newman, Inc.

Gordon, V. S., and Bieman, J. M. (1995). Rapid prototyping: Lessons learned. *IEEE Software*, 12(1), 85.

Gore, B. F. (2002). Human performance cognitive-behavioral modeling: A benefit for occupational safety. *International Journal of Occupational Safety and Ergonomics*, 8(3), 339–351.

Gore, B. F., Hooey, B. L., Foyle, D. C., and Scott-Nash, S. (2008). Meeting the challenge of cognitive human performance model interpretability though transparency: MIDAS v5. x. In *The 2nd International Conference on Applied Human Factors and Ergonomics*, July, pp. 14–17.

Gore, B. F., and Milgram, P. (2013). A validation approach for complex nextgen air traffic control human performance models. *Proceedings of the International Conference on Digital Human Modeling and Applications in Health, Safety, Ergonomics and Risk Management*, pp. 28–37.

Gore, B. F., and Smith, J. D. (2006). Risk assessment and human performance modelling: The need for an integrated systems approach. *International Journal of Human Factors Modelling and Simulation*, 1(1), 119–139.

Gregoriades, A., and Sutcliffe, A. (2008). Workload prediction for improved design and reliability of complex systems. *Reliability Engineering & System Safety*, 93(4), 530–549.

Hunn, B. P., and Heuckeroth, O. H. (2006). A shadow unmanned aerial vehicle (uav) improved performance research integration tool (IMPRINT) model supporting future combat systems. (Report no. ARL-TR-3731.) Aberdeen Proving Ground, MD: US Army Research Laboratory.

Jiang, X., Gao, Q., and Li, Z. (2013). Introducing human performance modeling in digital nuclear power industry. *Proceedings of the International Conference on Cross-Cultural Design*, pp. 27–36.

Kieras, D. E., and Meyer, D. E. (1995). Predicting human performance in dual-task tracking and decision making with computational models using the EPIC architecture. *Proceedings of the First International Symposium on Command and Control Research and Technology, National Defense University, June*. Washington, DC: National Defense University.

Kieras, D. E., and Meyer, D. E. (2000). The role of cognitive task analysis in the application of predictive models of human perfomance. In J. M. Schraagen, S. F. Chipman, and V. L. Shalin, (eds.), *Cognitive Task Analysis*, pp. 237–260. Mahwah, NJ: Lawrence Erlbaum Associates.

Kieras, D. E., Wakefield, G. H., Thompson, E. R., Iyer, N., and Simpson, B. D. (2016). Modeling two-channel speech processing with the EPIC cognitive architecture. *Topics in Cognitive Science*, 8(1), 291–304.

Krausman, A. S., Elliott, L. R., Redden, E. S., and Petrov, P. (2005). Effects of visual, auditory, and tactile cues on army platoon leader decision making. *Proceedings of the 10th International Symposium Command and Control Research and Technology (ICCRTS '05)*, pp. 2–8.

Laird, J. E., Newell, A., and Rosenbloom, P. S. (1987). SOAR: An architecture for general intelligence. *Artificial Intelligence*, 33(1), 1–64.

Laughery, K. R., Plott, B., Matessa, M., Archer, S., and Lebiere, C. (2012). Modeling human performance in complex systems. In G. Salvendy (ed.), *Handbook of Human Factors and Ergonomics* (4th ed.), pp. 931–961. Hoboken, NJ: John Wiley.

Laughery, R., Archer, S., Plott, B., and Dahn, D. (2000). Task network modeling and the micro saint family of tools. *Proceedings of the Human Factors and Ergonomics Society Annual Meeting*, 44(6), 721–724.

Li, N., Kong, H., Ma, Y., Gong, G., and Huai, W. (2016). Human performance modeling for manufacturing based on an improved KNN algorithm. *The International Journal of Advanced Manufacturing Technology*, 84(1–4), 473–483.

Meyer, D. E., Glass, J. M., Mueller, S. T., Seymour, T. L., and Kieras, D. E. (2001). Executive-process interactive control: A unified computational theory for answering 20 questions (and more) about cognitive ageing. *European Journal of Cognitive Psychology*, 13(1–2), 123–164.

Mitchell, D. K. (2000). Mental workload and ARL workload modeling tools. (Technical note ARL-TN-161.) Aberdeen Proving Ground, MD: US Army Research Laboratory.

Mitchell, D. K. (2003). Advanced improved performance research integration tool (IMPRINT) vetronics technology test bed model development. (Report no. ARL-TN-0208.) Aberdeen Proving Ground, MD: US Army Research Laboratory.

Nan, C., and Sansavini, G. (2016). Developing an agent-based hierarchical modeling approach to assess human performance of infrastructure systems. *International Journal of Industrial Ergonomics*, 53, 340–354.

Oyewole, S. A., and Haight, J. M. (2011). Determination of optimal paths to task goals using expert system based on GOMS model. *Computers in Human Behavior*, 27(2), 823–833.

Paik, J., and Pirolli, P. (2015). ACT-R models of information foraging in geospatial intelligence tasks. *Computational and Mathematical Organization Theory*, 21(3), 274–295.

Pew, R. W. (2008). More than 50 years of history and accomplishments in human performance model development. *Human Factors*, 50(3), 489–496.

Ritter, F., and Kim, J. (2015). ACT-R: Frequently asked questions. http://acs.ist.psu.edu/projects/act-r-faq/act-r-faq.html#G1 (accessed September 20, 2016).

Rothrock, L., and Narayanan, S. (2011). *Human-In-the-Loop Simulations*. New York: Springer.

Salmon, P., Jenkins, D., Stanton, N., and Walker, G. (2010). Hierarchical task analysis vs. cognitive work analysis: Comparison of theory, methodology and contribution to system design. *Theoretical Issues in Ergonomics Science*, 11(6), 504–531.

Sebok, A., Wickens, C., and Sargent, R. (2015). Development, testing, and validation of a model-based tool to predict operator responses in unexpected workload transitions. *Proceedings of the Human Factors and Ergonomics Society Annual Meeting*, 59(1), 622–626.

Stanton, N. A., Salmon, P. M., Walker, G. H., Baber, C., and Jenkins, D. P. (2010). *Human Factors Methods: A Practical Guide for Engineering and Design*. Aldershot, UK: Ashgate Publishing.

Staveland, L. (1994). Man–machine integration design and analysis system (MIDAS) task loading model (TLM) experimental and software detailed design report. Moffet Field, CA: Ames Research Center. Contract NAS2-13210.

Tyler, S. W., Neukom, C., Logan, M., and Shively, J. (1998). The MIDAS human performance model. *Proceedings of the Human Factors and Ergonomics Society Annual Meeting*, 42(3), 320–324.

Wickens, C. D. (2008). Multiple resources and mental workload. *Human Factors*, 50(3), 449–455.

Wickens, C. D., McCarley, J. S., Alexander, A. L., Thomas, L. C., Ambinder, M., and Zheng, S. (2008). Attention–Situation Awareness (A-SA) model of pilot error. *Human Performance Modeling in Aviation*, 213–239.

Wickens, C. D., Sebok, A., Li, H., Sarter, N., and Gacy, A. M. (2015). Using modeling and simulation to predict operator performance and automation-induced complacency with robotic automation: A case study and empirical validation. *Human Factors*, 57(6), 959–975.

Zhong, S., Ma, H., Zhou, L., Wang, X., Ma, S., and Jia, N. (2012). Guidance compliance behavior on VMS based on SOAR cognitive architecture. *Mathematical Problems in Engineering, 2012*. Hindawi Publishing Corporation, Article ID 530561, doi:10.1155/2012/530561.

chapter eight

*Functional interactions of work**

This chapter hypothesizes that the semantic relationship between modules' functional descriptions is correlated with the functional interaction between the modules. A deeper comprehension of the functional interactions between modules enables designers to integrate complex systems during the early stages of the product design process. Existing approaches that measure functional interactions between modules rely on the manual provision of designers' expert analyses, which may be time consuming and costly. The increased quantity and complexity of products in the twenty-first century further exacerbates these challenges. This work proposes an approach to automatically quantify the functional interactions between modules, based on their textual technical descriptions. Compared with manual analyses by design experts who use traditional Design Structure Matrix approaches, the text mining driven methodology discovers similar functional interactions, while maintaining comparable accuracies. The case study presented in this work analyzes an automotive climate control system and compares the functional interaction solutions achieved by a traditional design team with those achieved following the methodology outlined in this chapter.

Introduction

To be successful in today's global market, companies try to offer competitive and highly differentiated products by analyzing and developing product functions that satisfy customers' needs (Umeda et al., 2005). A product's *function* represents its operational purpose that meets customers' requirements (Umeda et al., 2005). This high-level product function, which directly interfaces with the customer, can be composed of multiple modules that perform each subfunction. A module performs a specific function by controlling the interactions of the functions of components (Jose and Tollenaere, 2005). Analyzing a product's functional characteristics is the initial step of the design process and precedes the definition of other aspects such the form and material (Bohm and Stone, 2004; Bryant et al., 2005; Stone et al., 1999; Umeda et al., 2005). Therefore, engineering

* Reprinted with permission from Kang, S. W., and Tucker, C., An automated approach to quantifying functional interactions by mining large-scale product specification data, *Journal of Engineering Design*, 27(1–3), 2016.

designers need to understand both the functional interactions between each module and how these interactions impact the overall product. These functional interactions indicate the degree of modularity among the attached modules and enable the designers to create new modules for next generation products by integrating/maintaining current modules within a product family/product portfolio (Dahmus et al., 2001; Gershenson et al., 2003; Schilling, 2000). To design a product, the designers must analyze the degree of functional interactions between modules based on their expertise/domain knowledge (Browning, 2001; Danilovic and Browning, 2007; Helmer et al., 2010; Pimmler and Eppinger, 1994; Sharman and Yassine, 2004; Sosa et al., 2003, 2004; Yassine and Braha, 2003). However, expert manual analyses (e.g., analyses by designers that quantify functional interactions between modules) may be a time consuming and costly process (Liang et al., 2008; Mudambi and Schuff, 2010; Rockwell et al., 2008; Yanhong and Runhua, 2007; Yoon and Park, 2004). These challenges are further exacerbated by the constant increase in product quantity and complexity, that are primarily driven by customers' increasing desires for customizable products (Alizon et al., 2009; Christensen et al., 2005; Tucker and Kang, 2012; Tucker and Kim, 2011). For example, at the start of the twentieth century, 92 modules were required to construct a complete car. However, more than 3500 modules currently exist in a modern day vehicle (Ford Motor Company, 1989; Groote, 2005). Over 30,000 new consumer products launched into the market each year (Christensen et al., 2005). Therefore, the complexity of managing modular product designs and their inherent functional interactions becomes cumbersome (Dahmus et al., 2001).

This work measures the functional interactions between modules by analyzing the semantic relationships between the modules' functional descriptions. The methodology presented in this work quantifies the functional interactions between modules by employing text mining algorithms that analyze modules' functional descriptions, represented textually through technical manuals pertaining to each module. The authors of this work hypothesize that the semantic relationship between modules' functional descriptions is correlated with the functional interaction between the modules. A statistical analysis that compares the results of the text mining methodology with experts' manual analysis of functional interactions (Browning, 2001; Pimmler and Eppinger, 1994; Pimmler, 1994) is presented. This text mining driven methodology achieves results in a timely and efficient manner that are comparable to the designers' manual analyses.

This chapter is organized as follows. This section discusses the research motivation; Section 2 describes works related to the research; Section 3 outlines the proposed methodology; Section 4 presents an automotive climate system case study that demonstrates the feasibility of the methodology; the results of the case study are discussed in Section 5; and Section 6 provides the conclusion and future related work.

Related works

The literature review begins by discussing the functional model used during the early stages of the product development process (Section 2.1). Then, the literature regarding semantic analyses in the engineering design fields is presented (Section 2.2). Section 2.3 reviews the formation of a module on the basis of functional interactions between modules. In Section 2.4, the literature related to manual approaches for measuring functional interactions articulates the need for an automated methodology that analyzes these interactions.

Functional modeling in engineering design

A functional model in engineering design is a structured representation of standardized functions and the flows between these functions within the formalized design space. The functional model generates a chain of functions as a process connected by energy, signal, and material flows—the essential requirements for operating each function, hereby developing a conceptual product design (Baxter et al., 1994; Bonjour et al., 2009; Hirtz et al., 2002; Kurtoglu et al., 2009; Stone and Wood, 2000). The functions and flows are defined on a functional basis, which designers describe with standardized technical terminology (Stone and Wood, 2000).

In product design, functional modeling is a crucial step in defining a product's architecture, wherein the architecture indicates the product's functional structure through which the product's function is allocated to physical modules. Designers have created a functional architecture for a next generation product on the basis of a functional model and the functional interactions between the candidate modules (Kurtoglu et al., 2009; Sangelkar and McAdams, 2013; Sen et al., 2010). A quantitative functional model that captures both the product functionality and customer requirements has been proposed (Stone et al., 1999). This model employs modular theory and requires an assessment tool to build a product repository for grouping products based on functionality and customer requirements.

Analyzing the interdependency of a product's functions can be performed independent of the analysis of a product's form (Kurtoglu et al., 2009; McAdams et al., 1999). This allows designers to make an explicit connection between modules by measuring the interdependency of functions. These connections are usually inherent in the product development process. The functional model enables product development and manufacturing on a mass customized scale (Kahn et al., 2005).

A function dividing process (FDP) has been proposed to obtain subsystem level (e.g., module) functions from system level (e.g., product) functions (Taura and Nagai, 2013). A FDP divides a system level function in two ways: the decomposition-based dividing process and the casual-connection-based dividing process. The causal-connection-based dividing process addresses the functional interdependency among a group of modules, wherein the functions of a module are similar to the adjacent functions of a product at the same level. The decomposition-based dividing process addresses functional independency when the product function is realized with the functions of each independent module. The model generates new module functions by integrating existing sub system functions with a functional model.

In the context of systems engineering, the need for a standard design process has arisen due to the international trade of system products and services. Therefore, EIA 632 has been established for standardizing functions of systems (Martin, 2000). The ISO/IEC 15288 standard has been introduced in the engineering design community to provide a common/comprehensive design framework for managing system development projects (Arnold and Lawson, 2004). However, as the quantity and diversity of modules continue to increase, designers are presented with the challenge of searching through the entire design space for novel design knowledge that can lead to next generation products.

To overcome these challenges, the methodology presented in this chapter employs semantic analysis techniques that discover functions from a large set of text data sets that describe modules. Functional interactions between modules are then automatically quantified in order to aid designers during next generation product design by informing them of the modules that are tightly or loosely coupled.

Knowledge extraction via semantic analysis

Semantic analysis techniques that discover knowledge from large-scale textual data sets, have been proposed across a wide range of science and engineering disciplines. The utilization of these techniques in the science and engineering fields enables researchers to access an immense number of textual data sets by mining statistically significant terms. For example, Bollen et al. have predicted changes in the stock market by analyzing the moods inherent in large-scale Twitter feeds (Bollen et al., 2011). Eysenbach has quantified the social impact of scholarly articles based on a semantic analysis of buzz on Twitter (Eysenbach, 2011). Asur and Huberman have successfully predicted box office revenues by mining information from tweets (Asur and Huberman, 2010). Ginsberg et al. have presented a methodology that tracks influenza epidemics in a population by analyzing a large number of Google search queries that were semantically related to the term "influenza" (Ginsberg et al., 2009). Huang et al. identify complex phenotypes and demonstrate a disease–drug connectivity map by analyzing semantic relationships across multiple diseases query expression profiles (H. Huang et al., 2010). Tuarob et al. have retrieved health-related information from social media data by analyzing the semantics of heterogeneous features (Tuarob et al., 2013). Paul and Dredze were able to identify ailments along with the terms that represent the symptoms of each ailment by mining public health information from social network services such as Twitter (Paul and Dredze, 2014). These studies demonstrate that it is possible to automatically discover semantic information that attains statistically significant correlations with ground truth data, despite using minimal human supervision.

In the engineering design fields, semantic analysis techniques have been employed to extract design knowledge from text based product data, including customer feedback and product technical descriptions, in order to design products that better meet customers' needs (Ghani et al., 2006; Menon et al., 2003; Romanowski and Nagi, 2004). Researchers have extracted product functions from their functional descriptions, such as patents or official manuals, through text mining techniques (Kang et al., 2013; Tucker and Kang, 2012; Ghani et al., 2006; Tseng et al., 2007; Tuarob and Tucker, 2014). Menon et al. employed a vector space document representation technique to derive useful product development information from customer reviews (Menon et al., 2003). Tuarob and Tucker proposed the method to identify lead users and product demand by mining product attributes from a large scale of social media networks (Tuarob and Tucker, 2015a,b). Zhou et al. have extracted latent customer needs from customer product reviews through semantic analysis, which identifies the hidden analogical reasoning of customers' preferences (Zhou et al., 2015). Gu et al. have employed a semantic reasoning tool to represent functional knowledge as function-cell pairs, where the cell is defined as a conceptual structure denoting the structure category that interacts with similar functions (Gu et al., 2012). Ghani et al. have extracted semantics as product attributes on the basis of textual product descriptions by employing a generative model with the expectation maximization (EM) technique (Ghani et al., 2006). Tucker and Kang have extracted semantics as functions and behaviors of products from textual descriptions in order to discover cross-domain knowledge among multiple product domains (Tucker and Kang, 2012). Kang et al. have employed a text mining technique that quantifies functional similarities between end-of-life (EOL) product descriptions (Kang et al., 2013). Product modules that exhibited low functional similarity were deemed strong candidates for EOL value creation. These modules were then combined to form a product with enhanced functional capabilities made from EOL products.

Existing semantic-based techniques in the engineering literature have focused on employing text mining techniques either at the early stages of customer needs analyses or at the end-of-life decision-making stages. This work hypothesizes that semantic relationships between modules' functional descriptions are correlated with functional interactions between the modules. This work demonstrates the feasibility of employing text mining algorithms to automatically quantify the functions of a module and measure the functional correlations between modules. To quantify the functional interactions of each module, an automatic interaction measurement is proposed in this work that extracts functions and converts those functions into a vector space on the basis of the semantic relationships of each module.

Modularity based on functional interaction

Modules that make up a product interact with each other through relevant engineering principles and knowledge (Batallas and Yassine, 2006; Cascini and Russo, 2007; Hong and Park, 2014; Jiao et al., 2007). Each interaction type (material, energy, information, spatial) between modules allows the product to operate its functions in the correct manner (Sharman and Yassine, 2004). Therefore, engineers assemble a product with modules that share similar materials, energies, information, and even shapes. Designers describe a product's mechanisms in its official descriptions (e.g., patent, textbook, manual) with engineering terms or conceptually similar terms that indicate the interactions and operation processes (Eckert et al., 2005; Kim et al., 2006; Murphy et al., 2014). Hence, the official textual descriptions of modules contain its functional descriptions and knowledge about the potential interactions between other modules.

The research on modularity is derived from Suh's design axiom, which establishes an understanding of the interactions between modules (Suh, 1984). A module is a group of components having strong functional interactions that proceed to perform a specific function (Fujita and Ishii, 1997; Gershenson et al., 2003; Jose and Tollenaere, 2005). It is difficult to integrate a module with other modules existing within a product when there are no functional interactions among them (Sharman and Yassine, 2004). The definition of modularity is further revised by considering interdependent/independent interactions between modules (Gershenson et al., 2003; Gershenson et al., 2004). A module represents a unit of a product that independently performs a specific function. Product design methodologies based on modularity use modules as the standard units to construct products to increase the efficiency of both the product design and manufacturing processes (C. C. Huang and Kusiak, 1998). For example, Volkswagen utilizes a platform which is modularized with a floor panel, chassis, and so on, for their products (Thevenot and Simpson, 2006). The Ford Motor Company produces a climate control system to provide both heating and cooling for their customers by integrating 16 different modules (Pimmler and Eppinger, 1994). Huang et al. designed a digital circuit module containing end users' needs as functions by integrating electric components (C. C. Huang and Kusiak, 1998). Modules constitute a product and its functional interactions that achieve the product's primary function. Understanding the functional interactions between modules enables designers to identify which modules can be integrated when creating new modules or enhancing existing ones (Dahmus et al., 2001; Gershenson et al., 2003; Schilling, 2000). A modular product is constructed with multiple modules along with their functional interactions, as shown in Figure 8.1.

Each module performs a specific function(s), and these modules are connected to each other with different levels of functional interactions, as shown in Figure 8.1. Although the identification of a functional interaction is an important factor for developing both modules and products, quantifying the degree of functional interaction (e.g., the line thickness in Figure 8.1) between modules has, until now, primarily relied extensively on manual feedback, which can be costly and time consuming, especially as products become more complex.

The methodology presented in this work automatically quantifies the degree of functional interaction of each module on the basis of each module's functional description. This work supports designers by creating an automated algorithm that discovers which modules can be integrated into a new module or which modules are not suitable for functional modifications.

Functional interaction analysis through a design structure matrix

In the context of engineering design, many researchers have employed a design structure matrix to visualize the functional interactions among modules on the basis of experts' knowledge (Dahmus et al., 2001; J. Jiao et al.,

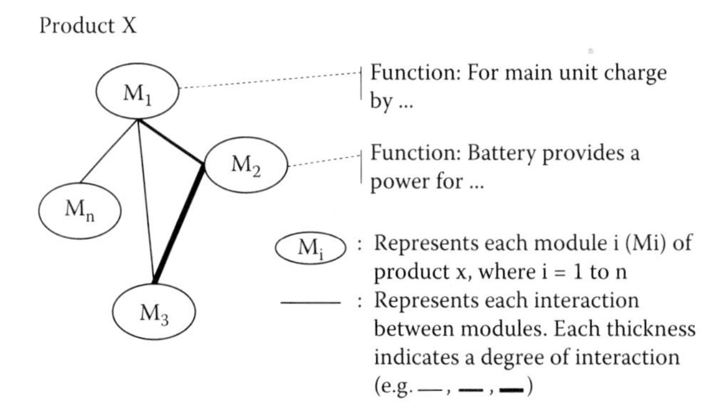

Figure 8.1 A modular product consisting of modules.

2007; Pimmler and Eppinger, 1994; Sosa et al., 2003, 2004). Steward coined the term "Design Structure Matrix (DSM)," a matrix-based approach to analyzing system design structures (Steward, 1981). The DSM's row index and the column index are described as system elements represented as modules, with the cells of the matrix representing the interactions between the elements. The initial stage of creating both matrices defines each *element* that is the object of interaction. Pimmler represented these system elements as product modules, proposing a taxonomy of functional interactions (spatial, energy, information, material) with a quantification scheme to facilitate the means by which experts measure each interaction between modules (Pimmler, 1994). Pimmler and Eppinger proposed a module-based DSM taxonomy to reorganize design teams along the lines of the functional interactions between modules (Pimmler and Eppinger, 1994). Eppinger extended the DSM to integrate manufacturing systems by mapping both the functional interactions of power train modules and the interactions of their manufacturing processes (Eppinger, 1997). Sosa et al. identified new modular and integrative systems to develop complex products by clustering the quantified interface between modules and systems (Sosa et al., 2003). They extended the research to analyze the functional misalignments of the product architecture and the organizational structure in complex product development by clustering the functional alignment of modules (Sosa et al., 2004). Karniel and Reich proposed a "DSM net" technique as a multilevel process model to create new products (Karniel and Reich, 2012). The DSM net is composed of design and process activities capable of checking process implementations on the basis of the functional interactions between modules. Existing DSM methodologies that measure functional interactions between modules are based on experts' analyses. However, it may be difficult for these manual analyses-based approaches to quantify the functional interactions between modules because modern engineering products, such as vehicles, require more complicated modules than did earlier products. For example, approximately 3400 more modules are required for constructing vehicles today than at the start of the twentieth century (Ford Motor Company, 1989; Groote, 2005). The methodology presented in this work automatically quantifies the functional interactions across a wide range of modules, thereby reducing the time and costs associated with manual analysis techniques.

In the early stages of the product design and development process, designers have been supported by automatic approaches or platforms in each step, as shown in Figure 8.2. Although designers are supported in each step by automated approaches or platforms, Step 5 is still heavily reliant on manual processes. The objective of Step 5 is to quantify the functional interactions between candidate modules for selecting appropriate modules that satisfy both the design specifications and their interactions.

Step 1. Identify customer requirements	Step 2. Establish target requirements	Step 3. Mapping requirements to design specifications	Step 4. Develop functional architecture for a product	Step 5. Measure functional associations
– Voice of customer (VOC)/needs extraction models based on sentiment analysis	– Consumer preference models – Customer opinion mining models – Discrete choice analysis (DCA)	– Quality function deployment (QFD) – Conjoint analysis – Convariance structural equation models – Nested clustering and aggregation models	– Functional model – ISO/IEC 15288 – EIA 632	Manual approaches have been employed on the basis of an expert's analysis, such as design structure matrix
Gamon et al., Yang et al., Wei et al., Zhan et al.	Petiot and Grognet, Hoyle et al., Malen and Hancock	Jiao et al., Huang and Mak, Zhang et al.	Bryant et al., Arnold and Lawson, Martin, Stone et al.	Primmler and Eppinger, Helmer et al.

Figure 8.2 The beginning phase of the new product development process. (From Arnold, S., and Lawson, H. W., *Systems Engineering*, *7*(3), 229–242, 2004; Bryant, C., et al., *International Conference on Engineering Design (ICED)* (Vol. 5), Melbourne, Australia, 2005; Gamon, M., et al., *Proceedings of the 6th International Conference on Advances in Intelligent Data Analysis (IDA '05)* (pp. 121–132), Springer, Berlin, 2005; Helmer, R., et al., *Journal of Engineering Design*, *21*(6), 647–675, 2010; Huang, G. Q., and Mak, K. L., *Journal of Engineering Design*, *10*(2), 183–194, 1999; Jiao, J. R., and Chen, C., *Concurrent Engineering*, *14*(3), 1–25, 2006; Martin, J. N., *Systems Engineering*, *3*(1), 2000; Petiot, J.-F., and Grognet, S., *Journal of Engineering Design*, *17*(3), 217–233, 2006; Pimmler, T. U., and Eppinger, S. D., *Proceedings of the 1994 ASME Design Theory and Methodology Conference (DTM '94)* (pp. 342–351), 1994; Stone, R. B., et al., *Journal of Engineering Design*, *19*(6), 489–514, 2008; Wei, C.-P., et al., *Information Systems and e-Business Management*, *8*(2), 149–167, 2009; Yang, C.-S., et al., *Proceedings of the 11th International Conference on Electronic Commerce* (pp. 64–71), 2009; Zhan, J., et al., *Expert Systems with Applications*, *36*(2), 2107–2115, 2009; Zhang, X. (Luke), et al., *Journal of Engineering Design*, *23*(1), 23–47, 2012.)

The methodology presented in Section 3 employs a topic model algorithm and a cosine measure to quantify the functional interactions among modules, with the purpose of moving toward automated methods that help to minimize the manual processes of Step 5 in the engineering design process (Figure 8.2).

Methodology

The methodology quantifies the degree of functional interactions of each module based on the module's functional descriptions by employing a topic model technique from natural language processing, thereby enabling designers to automatically model a product's functional architecture, and, as a result, minimize the manual analyses. This entire process, outlined in Figure 8.3, is defined as the automatic interaction measurement (AIM).

Step 1 describes the function data acquisition process for creating a database containing products' module information. In this work, the module descriptions are assumed to be represented textually. Each function extracted from a module's functional description is converted into a vector space in Step 2. Step 3 quantifies the functional interactions between the modules by measuring their vector similarities. Finally, the methodology can automatically measure the degree of functional interactions between modules and can serve as a guide for designers aiming to understand the complexities of the functional interactions within modular products.

Create a database of modules

The first step in the methodology is to construct a database that consists of a product's modules and contains its functional descriptions. A module's function data can be acquired from textual descriptions, such as patents or official manuals (Brunetti and Golob, 2000; Sheldon, 2009; Sheremetyeva et al., 1996; Umeda et al., 2005), as shown in Figure 8.4. It is assumed that each module has a unique identification number (ID) that will be used to

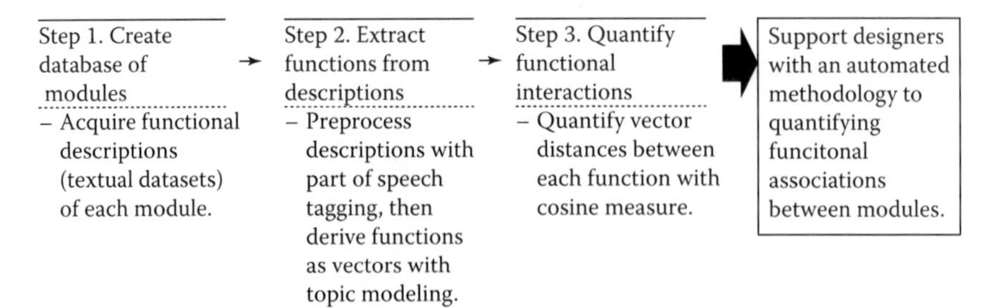
Step 1. Create database of modules
– Acquire functional descriptions (textual datasets) of each module.
Step 2. Extract functions from descriptions
– Preprocess descriptions with part of speech tagging, then derive functions as vectors with topic modeling.
Step 3. Quantify functional interactions
– Quantify vector distances between each function with cosine measure.
Support designers with an automated methodology to quantifying funcitonal associations between modules.

Figure 8.3 Flow diagram of the AIM.

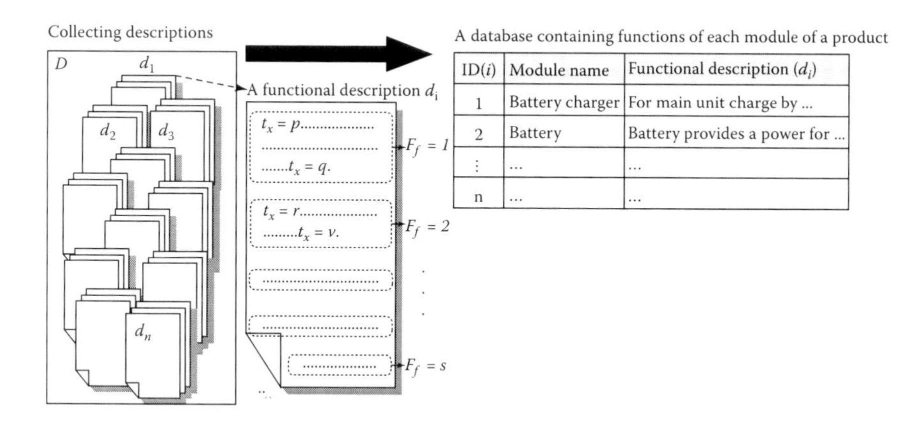

Figure 8.4 Process for extracting the functional descriptions from textual data.

automatically search for and discover the functional interactions between the modules.

- t_x represents the textual terms of a product's functional description (D).
- D represents a product's functional description, which is composed of the functional description of each module (e.g., $D = \{d_1, d_2, \ldots, d_n\}$).
- d_i represents each module's functional description.
- F_f is the fth paragraph of a functional description (d_i).
- S represents the total number of functions of the module that has the most functions.

A product's database composed of each module's functional descriptions is created, as shown in Figure 8.4. Because engineering documents such as a functional description of a module are laid out in a structured way, a paragraph of the description may contain one topic, which can be regarded as a function in this research (Nagle, 1996). The next section describes how the text mining algorithm extracts the functions from the modules' textual descriptions.

Extracting functions based on topic modeling

In the context of engineering design, official specifications, technical manuals, and patents can be regarded as the functional descriptions that include topics that can be represented as functions, as shown in Figure 8.4. Because a functional description is usually written in natural language, it includes many terms that do not provide any important information. Therefore, unnecessary terms such as linking verbs (e.g., is, and, etc.) are eliminated to reduce noise (Munková Daša et al., 2014;

Murphy et al., 2014). To extract functions from the descriptions, this work employs natural language processing techniques: a part-of-speech tagger and the latent Dirichlet allocation algorithm (Blei et al., 2003). Because functional terms mainly use verbs and nouns in functional descriptions, a part-of-speech (POS) tagger algorithm, which identifies verbs and nouns, is employed in this work (Ahmed et al., 2007; Toutanova and Manning, 2000). Once the POS tagger preprocesses the functional descriptions, latent Dirichlet allocation (LDA) extracts topics (e.g., functions) from the descriptions. LDA is a generative probabilistic model for compilations of text corpora, which can be regarded as functional descriptions with infinite mixtures over intrinsic topic groups (Blei et al., 2003). Because each paragraph of the description may describe each product's function with terms, the LDA algorithm postulates that the description is a finite mixture of the number of functions and that each term's establishment is due to one of the functions from the description (Sheldon, 2009). LDA provides the mixing proportions of functions through a generative probabilistic model on the basis of the Dirichlet distribution, as shown in the following Equation 8.1:

$$p(t_x|d_i) = \sum_{f-1}^{S} p(t_x|F_f)\, p(F_f|d_i) \tag{8.1}$$

where:

- t_x represents the textual terms of a product's functional description $D = \{d_1, d_2, \ldots, d_n\}$
- d_i represents each module's functional description
- F_f is the *f*th function of a product's functional description (D)
- S represents the total number of functions of the module with the most functions. The total number of functions (S) is the same as the number of paragraphs (specifications) in the functional description of the module that has the most functions.

From the set of modules of a product, each function is sequentially extracted with topic probabilities from the entire functional description ($D = \{d_1, d_2, \ldots, d_n\}$), as shown in Table 8.1—that is, Function 1 $p(F_1|D), \ldots$, Function S $p(F_S|D)$. These functions represent abstracts of each functional description in terms of contextual semantics. From Equation 8.1, $p(F_f|d_i)$ measures the probability of function (F_f) being a topic of a technical description (d_i). To compare the functional interaction between modules, each functional probability of the ith module ($p(F_f|d_i)$) represents the functional descriptive vector in matrix form, as shown in Table 8.1.

Table 8.1 Quantified functional data set

ID(i)	Module name	Function 1	Function 2	...	Function S
1	Battery charger	$p(F_1 \mid d_1)$	$p(F_2 \mid d_1)$	...	$p(F_S \mid d_1)$
2	Battery	$p(F_1 \mid d_2)$	$p(F_2 \mid d_2)$	...	$p(F_S \mid d_2)$
n	...	$p(F_1 \mid d_n)$	$p(F_2 \mid d_n)$	...	$p(F_S \mid d_n)$

Quantifying functional interactions

To search for functional interactions across textual data sets, the cosine measure has been employed using the LDA results. The cosine measure is employed in this work to quantify the degree of functional interaction between each module. The functional interaction between modules can be quantified by inputting the functional descriptive vectors from Table 8.1 into the cosine measure. For instance, the functional interaction between the functional vector "Battery charger" and "Battery" can be quantified with the following Equation 8.2:

$$\text{Cos}\left(V_{i=1} \middle| V_{i=2}\right) = \frac{V_{i=1} \cdot V_{i=2}}{V_{i=1}V_{i=2}} \tag{8.2}$$

where:

$$V_{i=1} \cdot V_{i=2} = \sum_{f=1}^{S} p\left(F_f \middle| d_1\right) p\left(F_f \middle| d_2\right) \tag{8.3}$$

$$V_{i=1} = \sqrt{\sum_{f=1}^{S} p\left(F_f \middle| d_1\right)^2} \tag{8.4}$$

$$V_{i=2} = \sqrt{\sum_{f=1}^{S} p\left(F_f \middle| d_2\right)^2} \tag{8.5}$$

Each variable $V_{i=1}$ and $V_{i=2}$ represents a vector coordinate of the *Battery charger's* ($i = 1$) and *Battery's* ($i = 2$) functions in Table 8.1. The cosine measure is 1 when the angle between the two vectors is 0 degrees, while the cosine measure is 0 when the angle between the two vectors is 90 degrees. Therefore, the functional interaction increases when the cosine metric between the functional descriptive vectors is close to 1, whereas 0 means that there are no interactions between the functions. Modules that have strong functional interactions with one another can be integrated into a new module, while modules with low functional interactions may be independently updated/enhanced with minimal impact on the other modules (Browning, 2001; Hirtz et al., 2002). In this work, if the cosine measure between the functional descriptive vectors results in a value of 1, it is assumed that the corresponding modules are identical. Sophisticated engineering products, ranging from automobiles to aircrafts, may include multiple identical modules in one system. For instance, most cars have two headlights that perform independently of one another. Because the interactions between one headlight and another module (e.g., wheel) would result in the same cosine similarity value, regardless of whether it is the right headlight or the left headlight, directly searching functional interactions across identical modules is redundant and therefore not considered in the methodology.

```
Input: d_i: a functional description for i^th module
        Set i = 1
        Loop
                Run POS
                Record POS result
                                If i > n
End; POS end
import record POS result (i = 1 to n)
                Set i = 1
                Loop
                                Set f = 1
                                                Loop
                                                Run LDA
                                                Record LDA result
                                                                Set f = f +1
                                                                If f > S
                                                End
                                Set i = i+1
                                If i > n
End; LDA end
Import record LDA result (i = 1 to n)
Set i = 1
                Loop
                Set j = i + 1
                                Loop
                                Run cosine measure (i,j)
                Set t = cosine measure results
                        If t < 1
                                                Print out t
                                Set j = j + 1
                                If j > n
                                End
                Set i = i + 1
                If i > n
End; cosine measure end
```

Figure 8.5 Algorithm flow of the AIM.

In the early stages of the product development process, designers consider functional interactions between different modules for functional architecture modeling. Figure 8.5 presents the algorithmic flow of the automated interaction measurement (AIM). The AIM imports data sets from a database and then cleanses the textual data by employing the POS tagger. LDA extracts functions from the POS results, and then the cosine measure quantifies the functional interactions across the modules on the basis of the LDA results.

In contrast to traditional DSM approaches, the methodology outlined in this work analyzes functional interactions in an automated manner, with minimal manual input from designers. This is particularly important as the number of modules and functional interactions increase in complex products. To evaluate the AIM, the next section introduces a DSM study as a case study for comparing the interaction analysis.

Application

The case study analyzes an automotive climate control system that combines 16 modules that have 120 functional interactions. The case study is introduced to verify the feasibility of the AIM presented in Section 3 by comparing it to manual analyses performed by design experts. Pimmler and Eppinger extracted functional interactions between the automotive climate control system's modules through a taxonomy of functional interactions and a manual quantification process, as shown in Table 8.2 (Pimmler and Eppinger, 1994). Their study analyzed modules with

Table 8.2 Interaction types and quantification of the DSM

Type	Interaction values
Spatial	Needs for adjacency or orientation between two modules. Required(+2)/desired(+1)/indifferent(0)/undesired(−1)/detrimental(−2)
Energy	Needs for energy transfer/exchange two modules. Required(+2)/desired(+1)/indifferent(0)/undesired(−1)/detrimental(−2)
Information	Needs for data or signal exchange between two modules. Required(+2)/desired(+1)/indifferent(0)/undesired(−1)/detrimental(−2)
Material	Needs for material exchange between two modules. Required(+2)/desired(+1)/indifferent(0)/undesired(−1)/detrimental(−2)

Source: Pimmler, T. U., and Eppinger, S. D., *Proceedings of the 1994 ASME Design Theory and Methodology Conference (DTM '94)* (pp. 342–351), The American Society of Mechanical Engineers, New York, NY, 1994.

four different interaction types (spatial, energy, information, material), based on five different scores (required/desired/indifferent/undesired/detrimental) (Pimmler and Eppinger, 1994). Functional interactions are quantified by four different generic relationship types with values of −2, −1, 0, 1, and 2, as shown in Table 8.2. In Pimmler and Eppinger's study, a design structure matrix was generated by conducting interviews with experts from the Ford Motor Company; the original DSM is described in the Figure 8.6 Appendix (Pimmler and Eppinger, 1994). Although the manual DSM analysis provides reliable outputs, it may be difficult to extract functional interactions, as the quantity and complexity of modules continues to increase in today's twenty-first-century product space.

	A	B	C	D	E	F	G	H	I	J	K	L	M	N	O	P
Radiator A		2 0 0 2			2 −2 0 0											
Engine fan B	2 0 0 2				2 0 0 2								1 0 0 0			
Heater core C				1 0 0 0			2 0 0 0	−1 0 0 0								0 0 0 2
Heater hoses D			1 0 0 0						−1 0 0 0							
Conedenser E	2 −2 0 0	2 0 0 2				0 2 0 2		−2 2 0 2								
Compressor F					0 2 0 2			0 2 0 2	1 0 0 2	0 0 2 0	0 0 2 0		1 0 0 0			
Evaporator case G			2 0 0 0					2 0 0 0						2 0 0 0	2 0 0 0	2 0 0 2
Evaporator core H			−1 0 0 0		−2 2 0 2	0 2 0 2	2 0 0 0		1 0 0 2							0 0 0 2
Accumulator I				−1 0 0 0		1 0 0 2		1 0 0 2		1 0 0 0						
Refrigeration controls J						0 0 2 0			1 0 0 0		0 0 2 0		1 0 0 0			
Air controls K						0 0 2 0				0 0 2 0		0 0 2 0	1 0 0 0	0 0 2 0	0 0 2 0	
Sensors L											0 0 2 0		1 0 0 0			
Commnad distribution M		1 0 0 0				1 0 0 0				1 0 0 0	1 0 0 0	1 0 0 0		1 0 0 0	1 0 0 0	1 0 0 0
Actuators N							2 0 0 0				0 0 2 0		1 0 0 0			
Blower controller O							2 0 0 0				0 0 2 0		1 0 0 0			2 0 0 2
Blower motor P			0 0 0 2				2 0 0 2	0 0 0 2					1 0 0 0		2 0 0 2	

Note: Blank matrix elements indicate no interaction (four zero scrores).

Legend:

Spatial:	S E	:Energy
Information:	I M	:Materials

Figure 8.6 Appendix. (From Pimmler, T. U., and Eppinger, S. D., *Proceedings of the 1994 ASME Design Theory and Methodology Conference (DTM '94)* (pp. 342–351), The American Society of Mechanical Engineers, New York, NY, 1994.)

The methodology presented in this work mitigates this challenge by automatically analyzing functional interactions on the basis of modules' functional descriptions. To verify the feasibility of the methodology, this case study compares the interactions from the AIM to those of the DSM generated by manual expert feedback that has been thoroughly studied in the engineering design community (Browning, 2001; Pimmler and Eppinger, 1994).

Sosa et al. have proposed the "design interface strength" to measure the degree of the overall functional interactions between modules on the basis of each interaction type and its scale value from Pimmler and Eppinger's research, shown in Table 8.2 (Sosa et al., 2003, 2004).

The interaction scale ranges from 0 (e.g., all values of each generic interaction are 0 or a negative value) to 8 (e.g., all values of each generic interaction are 2) by integrating the four generic interaction types: spatial $|2|$ + energy $|2|$ + information $|2|$ + material $|2|$ for a maximum of 8. To distinguish modules in the system, Sosa et al. categorized the strength of the functional interactions into "low" (less than 4) and "high" (greater than 4) (Sosa et al., 2004).

In this paper, because the interactions are measured on a {0, 1} scale, the interaction values of the DSM need to be normalized to the same scale. The manually analyzed interaction values of the automotive control system are normalized to a {0, 1} scale, as shown in Table 8.3.

In Table 8.3, "M" stands for the module, module 1 represents the radiator, module 2 represents the engine fan, module 3 represents the heater core, module 4 represents the heater hoses, module 5 represents the condenser, module 6 represents the compressor, module 7 represents the evaporator case, module 8 represents the evaporator core, module 9 represents the accumulator, module 10 represents the refrigerator controls, module 11 represents the air controls, module 12 represents the sensors, module 13 represents the command distribution, module 14 represents the actuator, module 15 represents the blower controller, and module 16 represents the blower motor.

Referring to Sosa et al., interactions can be divided into high or low values on the basis of the average value across the scale (Sosa et al., 2004). Therefore, each degree of interaction is divided into high (H) and low values (L) in this paper, based on whether a functional interaction is greater or less than 0.5 (given a scale from 0 to 1).

Table 8.4 describes the transformed functional interactions, which are based on the functional interaction degrees from Table 8.3. The automotive climate control system was analyzed by design experts, whereas this work automatically quantifies the degree of interactions by automatically analyzing the functional descriptions based on the LDA algorithm (Ramage and Rosen, 2009).

Table 8.3 Normalized DSM functional interactions on the basis of a manual analysis

Module	1	2	3	4	5	6	7	8	9	10	11	12	13	14	15	16
1		.50	.00	.00	.50	.00	.00	.00	.00	.00	.00	.00	.00	.00	.00	.00
2			.00	.00	.50	.00	.00	.00	.00	.00	.00	.00	.13	.00	.00	.00
3				.13	.00	.00	.25	.00	.00	.00	.00	.00	.00	.00	.00	.25
4					.00	.00	.00	.00	.00	.00	.00	.00	.00	.00	.00	.00
5						.50	.00	.50	.00	.00	.00	.00	.00	.00	.00	.00
6							.00	.50	.38	.25	.25	.00	.13	.00	.00	.00
7								.25	.00	.00	.00	.00	.00	.25	.25	.50
8									.38	.00	.00	.00	.00	.00	.00	.25
9										.13	.00	.00	.00	.00	.00	.00
10											.25	.00	.13	.00	.00	.00
11												.25	.13	.25	.25	.00
12													.13	.00	.00	.00
13														.13	.13	.13
14															.00	.00
15																.50
16																

Table 8.4 Transformed functional interactions on the basis of the normalized interactions

Module	1	2	3	4	5	6	7	8	9	10	11	12	13	14	15	16
1		H	L	L	H	L	L	L	L	L	L	L	L	L	L	L
2			L	L	H	L	L	L	L	L	L	L	L	L	L	L
3				L	L	L	L	L	L	L	L	L	L	L	L	L
4					L	L	L	L	L	L	L	L	L	L	L	L
5						H	L	H	L	L	L	L	L	L	L	L
6							L	H	L	L	L	L	L	L	L	L
7								L	L	L	L	L	L	L	L	H
8									L	L	L	L	L	L	L	L
9										L	L	L	L	L	L	L
10											L	L	L	L	L	L
11												L	L	L	L	L
12													L	L	L	L
13														L	L	L
14															L	L
15																H
16																

Create a database of the automotive climate control system's modules

To perform the experiment, this research follows each step of the AIM (referring to Figure 8.3 of Section 3). Each module's functional description has been collected from Daly's document, as shown in Figure 8.7 (Daly, 2006).

- D represents the automotive climate control system's functional description, which is composed of the functional description of each module (e.g., $D = \{d_1, d_2, \ldots, d_{16}\}$)
- d represents each module's functional description
- t_x represents the textual terms of the automotive climate control system's functional description (D)
- F_f is the fth paragraph of a functional description (d_i)
- S represents total number of functions of the module that has the most functions (e.g., S = 10 from d_8; the other descriptions have less than 10 paragraphs)

On the basis of the data collection process, the automotive climate control system's database, composed of each module's functional descriptions is created, as shown in Table 8.5.

Extracting functions from the functional descriptions

The collected descriptions have been preprocessed by the POS tagger to extract nouns and verbs before performing LDA (Table 8.6). Both the preprocessing and function extraction processes are taken into account in Step 2 of the methodology (Figure 8.3). These natural language processing techniques are based on the Stanford Natural Language Processing platform (Ramage and Rosen, 2009).

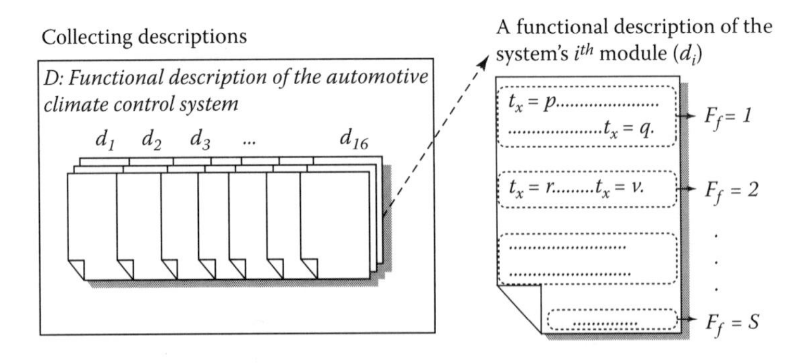

Figure 8.7 Functional description extraction process from textual data.

Table 8.5 A database containing functions of each module of the climate control system

ID(i)	Module	Functional Description (d_i)
1	Radiator	The radiator dissipates excess engine heat, …
2	Engine fan	The engine fan draws outside air into the engine, …
3	Heater core	The heater core transfers heat energy via forced, …
⋮	⋮	⋮
16	Blower motor	The blower motor moves fresh or vehicle interior air, …

Table 8.6 Preprocess functional descriptions by the POS tagger

ID (i = 1 to 16)	Module	Description Preprocess by the POS tagger (n: noun, v: verb)
1	Radiator	The radiator dissipates excess engine heat, … Radiator (n) dissipates (v) engine (n) heat (n), …
2	Engine fan	The engine fan draws outside air into the, … Engine fan (n) draws (v) air (n), …
⋮	⋮	⋮
16	Blower motor	The blower motor moves fresh air or, … Blower motor (n) moves (v) air (n), …

Table 8.7 Extracted functions from the automotive climate control system

	F_1 = heat	F_2 = absorbs	F_3 = transfers	F_4 = coolant	…	F_{10} = accumulator
$P(F_f \mid D)$	0.14257	0.09431	0.09209	0.08999	…	0.00421

The AIM presented in this work measures the degrees of functional interactions between modules on the basis of functional descriptions. Given the functional descriptions of the 16 modules, the evaporator core (module 8) has the most functions among all the modules: $S = 10$, referring to variable S in Equation 8.1.

Therefore, 10 functions (e.g., heat, absorbs, transfers, coolant, air, energy, provides, temperature, exchanger, accumulator) are extracted from the preprocessed automotive climate control system's entire functional description (D) by LDA, as shown in Table 8.7.

After each term representing the function of the climate control system is extracted, LDA quantifies the probabilities of each function being a topic of each module's technical description. Each functional probability ($p(F_f \mid d_i)$) of the climate control systems' modules represents the functional descriptive vector in matrix form, as shown in Table 8.8.

Quantifying the functional interactions between components is the final step (Step 3 in Figure 8.3 of Section 3) of the methodology presented

Table 8.8 Quantified functional data set for each module's description

ID ($i = 1$ to 16)	Module	F_1 = heat	F_2 = absorbs	F_3 = transfers	F_4 = coolant	...	F_{10} = accumulator
1	Radiator	0.03615	0.01503	0.09474	0		0.10043
2	Engine fan	0.01274	0.10572	0.08347	0	...	0.02443
⋮	⋮	⋮	⋮	⋮	⋮	⋮	⋮
16	Blower motor	0	0	0.00045	0.02899	...	0

in this research. The next section describes 120 interactions that are quantified by the AIM on the basis of the values from the LDA results. To verify the feasibility of the AIM, these interactions are compared with those of the manually analyzed DSM (Table 8.4) on the basis of the statistical verification models presented in Section 5.

Results and discussion

The manually analyzed DSM has been shown to be effective for analyzing interactions between objects by providing designers with valuable results. Therefore, comparable results of the manually analyzed DSM and the AIM presented in this work will demonstrate the feasibility of quantifying functional interactions in an automated manner so that designers can focus more on idea generation, rather than on functional mapping.

The degrees of the functional interactions are then quantified by the cosine measure (Equation 8.2). The quantified interactions of each module are presented in Table 8.9. Each module (i.e., 1 to 16) represents the same modules presented in Table 8.4.

The interaction values from the AIM are transformed to binary values (high or low) using the same scale as the manual process, as shown in Table 8.10. Because the functional interaction degrees (i.e., manual analysis and the AIM) have been transformed to the same scale, a paired T-test and a confusion matrix are generated to provide statistical evidence of the similarity of the results from the AIM (Table 8.10) and the manual analysis (Table 8.4).

Statistical verification: Paired T-test

Because a manual DSM has been shown to be effective for analyzing functional interactions, a paired T-test is performed to determine whether there is a statistically significant difference between the baseline results (e.g., manual DSM results) and the results generated by the AIM. This is achieved by comparing the degrees of functional interactions quantified using the proposed model (Table 8.10) and the manual DSM (Table 8.4).

Table 8.9 Functional interaction by the AIM

Module	1	2	3	4	5	6	7	8	9	10	11	12	13	14	15	16
1		.41	.47	.00	.75	.00	.00	.35	.00	.00	.00	.00	.14	.00	.00	.00
2			.00	.00	.22	.00	.08	.00	.00	.02	.28	.00	.00	.00	.01	.18
3				.28	.73	.40	.44	.93	.14	.01	.00	.00	.12	.00	.01	.28
4					.08	.00	.54	.03	.00	.01	.00	.00	.00	.00	.01	.00
5						.58	.08	.58	.21	.01	.01	.00	.13	.00	.01	.01
6							.00	.37	.36	.35	.00	.00	.00	.00	.01	.00
7								.41	.03	.21	.20	.00	.10	.40	.41	.50
8									.13	.01	.00	.00	.09	.00	.00	.39
9										.14	.00	.00	.00	.04	.01	.03
10											.01	.00	.01	.17	.42	.01
11												.32	.45	.00	.01	.08
12													.00	.00	.00	.00
13														.00	.36	.00
14															.34	.78
15																.01
16																

Table 8.10 Transformed functional interactions from the AIM results

Module	1	2	3	4	5	6	7	8	9	10	11	12	13	14	15	16
1		L	L	L	H	L	L	L	L	L	L	L	L	L	L	L
2			L	L	L	L	L	L	L	L	L	L	L	L	L	L
3				L	H	L	L	H	L	L	L	L	L	L	L	L
4					L	L	H	L	L	L	L	L	L	L	L	L
5						H	L	H	L	L	L	L	L	L	L	L
6							L	L	L	L	L	L	L	L	L	L
7								L	L	L	L	L	L	L	L	H
8									L	L	L	L	L	L	L	L
9										L	L	L	L	L	L	L
10											L	L	L	L	L	L
11												L	L	L	L	L
12													L	L	L	L
13														L	L	L
14															L	H
15																L
16																

The paired T-test's null hypothesis assumes that the mean difference of paired values is 0. The paired values are the degrees of the functional interactions of each module from the manual DSM (Table 8.4) and the AIM (Table 8.10) results. The values from each analysis are paired if they have the same indices in both tables (the row and column indices of Table 8.4 and Table 8.10 represent the same modules). The paired T-test is performed for 120 paired values (excluding values of identical row–column indices) to statistically determine whether the proposed automated and manual DSM results are significantly different. If the test does not reject the null hypothesis, the AIM can be regarded as a valid model for analyzing the functional interactions for this case study. Based on the paired T-test results (N = 120), the mean difference of the T-test is 0. The results of this analysis indicate that the null hypothesis is not rejected, with a T-value of 0.00, a P-value of 1.000, and an α of 0.05. Because the null hypothesis is strongly supported by having a P-value (1.000) greater than α (0.05), there is no significant difference between the functional interactions of each module from the proposed automated approach and the manual DSM. This test statistically verifies that the AIM provides functional interactions similar to the manually analyzed interactions.

Statistical verification: Confusion matrix

The confusion matrix (Table 8.11) shows that the AIM has 94% accuracy when benchmarked against the manual DSM generation; of a total of 120 instances, 114 (low: 110, high: 4) instances from the predictive model (the AIM) are matched with the actual model (DSM). In this case study, the AIM analyzes low interactions among the modules with 98% precision and 96% recall.

Low interaction can be regarded as a functional independence (modularity) that affects how designers construct a system architecture with unique modules. These modules can be assembled for serving their own independent functions within the system. However, the AIM analyzes high interactions among the modules with 50% precision and 67% recall, thereby providing designers with insufficient information regarding how modules should be integrated when creating a new module for next generation products. Although the AIM presented in this work performed less accurately for extracting high interactions between modules, it discovers

Table 8.11 Confusion matrix: DSM vs. the AIM

		Predictive class (the AIM)	
		L (Low)	H (High)
Actual class (DSM)	L (Low)	110	2
	H (High)	4	4

functionally detachable modules and guides designers in terms of which modules can be potentially detached, revised, or enhanced, with minimal impact on other subsystems.

A text mining technique may provide a more efficient means of quantifying functional interactions (especially for low functional interactions) between modules when compared with a manually generated DSM analysis, because the number of modules continues to increase along with their functional descriptions. To support designers with an analysis that is compatible with experts' manual analyses, the methodology needs to be improved for extracting high functional interactions in future work.

Conclusions and future work

This work includes an extensive literature review of research in the engineering design and text mining fields. The literature review has shown that engineering design methodologies continue to introduce more automated approaches that support designers at both the customer needs and end-of-life product stages of the design process. Semantic analyses have been employed in the engineering design field to discover design information from customer reviews and functional descriptions. This work hypothesizes that semantic relationships between modules' functional descriptions are correlated with functional interactions between the modules. To support designers in integrating/maintaining modules during the concept generation process, this work automatically measures the functional interactions between modules. By employing the LDA algorithm the cosine metric, the methodology presented in this work discovers functional interactions between modules on the basis of semantic relationships between textual data sets that describe the modules' functions. Furthermore, the AIM has been validated using a case study involving a DSM analysis of an automotive climate system. The case study is conducted on a limited data set. The results achieved indicate the methodology's working prospect scope. The authors would like to emphasize that this work explores a correlation (not a causal relationship) between modules' functional descriptions and modules' functional interactions.

The functional interactions between modules allow designers to efficiently create a product's architecture by integrating the modules' functions (Fixson, 2007; Gershenson et al., 2003; C. C. Huang and Kusiak, 1998; Sosa et al., 2003). Functional interactions typically indicate the degree of modularity among modules at the beginning of the product development process, thereby enabling designers to make decisions such as to extend, upgrade, or maintain existing modules. The AIM algorithm presented in this work performed less accurately for extracting high interactions between modules than deriving low interactions. Thus, improving the methodology to accurately extract high functional interactions from functional descriptions may enable designers to discover modules that can be integrated during the creation of new modules for next generation products.

References

Ahmed, S., Kim, S., and Wallace, K. M. (2007). A methodology for creating ontologies for engineering design. *Journal of Computing and Information Science in Engineering*, *7*(2), 132–140.

Alizon, F., Shooter, S. B., and Simpson, T. W. (2009). Assessing and improving commonality and diversity within a product family. *Research in Engineering Design*, *20*(4), 241–253.

Arnold, S., and Lawson, H. W. (2004). Viewing systems from a business management perspective: The ISO/IEC 15288 standard. *Systems Engineering*, *7*(3), 229–242.

Asur, S., and Huberman, B. A. (2010). Predicting the future with social media. In *Web Intelligence and Intelligent Agent Technology (WI-IAT), 2010 IEEE/WIC/ACM International Conference* (Vol. *1*, pp. 492–499). doi:10.1109/WI-IAT.2010.63

Batallas, D., and Yassine, A. (2006). Information leaders in product development organizational networks: Social network analysis of the design structure matrix. *IEEE Transactions on Engineering Management*, *53*(4), 570–582.

Baxter, J. E., Juster, N. P., and De Pennington, A. (1994). A functional data model for assemblies used to verify product design specifications. *Proceedings of the Institution of Mechanical Engineers, Part B: Journal of Engineering Manufacture*, *208*(4), 235–244.

Blei, D. M., Ng, A. Y., and Jordan, M. I. (2003). Latent Dirichlet allocation. *Journal of Machine Learning Research*, *3*, 993–1022.

Bohm, M. R., and Stone, R. B. (2004). Representing functionality to support reuse: Conceptual and supporting functions. In *ASME Design Engineering Technical Conference and Computers and Information in Engineering Conference* (pp. 411–419). Salt Lake City, UT. doi:10.1115/DETC2004-57693

Bollen, J., Mao, H., and Zeng, X. (2011). Twitter mood predicts the stock market. *Journal of Computational Science*, *2*(1), 1–8.

Bonjour, E., Deniaud, S., Dulmet, M., and Harmel, G. (2009). A fuzzy method for propagating functional architecture constraints to physical architecture. *Journal of Mechanical Design*, *131*(6). doi:10.1115/1.3116253

Browning, T. R. (2001). Applying the design structure matrix to system decomposition and integration problems: A review and new directions. *IEEE Transactions on Engineering Management*, *48*(3), 292–306.

Brunetti, G., and Golob, B. (2000). A feature-based approach towards an integrated product model including conceptual design information. *Computer-Aided Design*, *32*(14), 877–887.

Bryant, C., Stone, R., McAdams, D., Kurtoglu, T., and Campbell, M. (2005). Concept generation from the functional basis of design. In *International Conference on Engineering Design (ICED)* (Vol. *5*). Melbourne, Australia.

Cascini, G., and Russo, D. (2007). Computer-aided analysis of patents and search for TRIZ contradictions. *International Journal of Product Development*, *4*(1), 52–67.

Christensen, C. M., Cook, S., and Hall, T. (2005). Marketing malpractice: The cause and the cure. *Harvard Business Review*, *83*(12), 74–83. Retrieved from https://hbr.org/2005/12/marketing-malpractice-the-cause-and-the-cure.

Dahmus, J. B., Gonzalez-Zugasti, J. P., and Otto, K. N. (2001). Modular product architecture. *Design Studies*, *22*(5), 409–424.

Daly, S. (2006). *Automotive Air-Conditioning and Climate Control Systems*. Boston, MA: Butterworth-Heinemann.

Danilovic, M., and Browning, T. R. (2007). Managing complex product development projects with design structure matrices and domain mapping matrices. *International Journal of Project Management*, *25*(3), 300–314.

Eckert, C. M., Martin, S., and Christopher, E. (2005). References to past designs. In J. S. Gero and N. Bonnardel (eds.), *Studying Designers 5* (pp. 3–19). Sydney, Australia: Key Centre of Design Computing and Cognition.

Eppinger, S. D. (1997). A planning method for integration of large-scale engineering systems. In *Proceedings of the International Conference on Engineering Design (ICED)* (pp. 199–204). Tampere, Finland.

Eysenbach, G. (2011). Can tweets predict citations? Metrics of social impact based on Twitter and correlation with traditional metrics of scientific impact. *Journal of Medical Interest Research*, *13*(4). doi:10.2196/jmir.2012

Fixson, S. K. (2007). Modularity and commonality research: Past developments and future opportunities. *Concurrent Engineering*, *15*(2), 85–111.

Ford Motor Company. (1989). *1919–1927 Model T: Ford Service: Practical Methods for Repairing and Servicing Ford Cars*. Detroit, MI: Ford Motor Company.

Fujita, K., and Ishii, K. (1997). Task structuring toward computational approaches to product variety design. In *Proceedings of the ASME International Design Engineering Technical Conferences*. New York: ASME.

Gamon, M., Aue, A., Corston-Oliver, S., and Ringger, E. (2005). Pulse: Mining customer opinions from free text. In *Proceedings of the 6th International Conference on Advances in Intelligent Data Analysis (IDA '05)* (pp. 121–132). Berlin: Springer. doi:10.1007/11552253_12

Gershenson, J. K., Prasad, G. J., & Zhang, Y. (2003). Product modularity: Definitions and benefits. *Journal of Engineering Design*, *14*(3), 295–313.

Gershenson, J. K., Prasad, G. J., and Zhang, Y. (2004). Product modularity: Measures and design methods. *Journal of Engineering Design*, *15*(1), 33–51.

Ghani, R., Probst, K., Liu, Y., Krema, M., and Fano, A. (2006). Text mining for product attribute extraction. *SIGKDD Explorations Newsletter*, *8*(1), 41–48.

Ginsberg, J., Mohebbi, M. H., Patel, R. S., Brammer, L., Smolinski, M. S., and Brilliant, L. (2009). Detecting influenza epidemics using search engine query data. *Nature*, *457*(7232), 1012–1014.

Groote, S. De. (2005). Construction of an F1 car. Retrieved from http://www.f1technical.net/articles/1271.

Gu, C.-C., Hu, J., Peng, Y.-H., and Li, S. (2012). FCBS model for functional knowledge representation in conceptual design. *Journal of Engineering Design*, *23*(8), 577–596.

Helmer, R., Yassine, A. A., and Meier, C. (2010). Systematic module and interface definition using component design structure matrix. *Journal of Engineering Design*, *21*(6), 647–675.

Hirtz, J., Stone, R. B., McAdams, D. A., Szykman, S., and Wood, K. L. (2002). A functional basis for engineering design: Reconciling and evolving previous efforts. *Research in Engineering Design*, *13*(2), 65–82.

Hong, E.-P., and Park, G.-J. (2014). Modular design method based on simultaneous consideration of physical and functional relationships in the conceptual design stage. *Journal of Mechanical Science and Technology*, *28*(1), 223–235.

Huang, C. C., and Kusiak, A. (1998). Modularity in design of products and systems. *IEEE Transactions on Systems, Man and Cybernetics, Part A: Systems and Humans*, *28*(1), 66–77.

Huang, G. Q., and Mak, K. L. (1999). Web-based collaborative conceptual design. *Journal of Engineering Design*, *10*(2), 183–194.

Huang, H., Liu, C.-C., and Zhou, X. J. (2010). Bayesian approach to transforming public gene expression repositories into disease diagnosis databases. *Proceedings of the National Academy of Sciences of the United States of America*, *107*(15), 6823–6828.

Jiao, J. R., and Chen, C. (2006). Customer requirement management in product development: A review of research issues. *Concurrent Engineering*, *14*(3), 1–25.

Jiao, J., Simpson, T. W., and Siddique, Z. (2007). Product family design and platform-based product development: A state-of-the-art review. *Journal of Intelligent Manufacturing*, *18*(1), 5–29.

Jose, A., and Tollenaere, M. (2005). Modular and platform methods for product family design: Literature analysis. *Journal of Intelligent Manufacturing*, *16*(3), 371–390.

Kahn, K. B., Castellion, G., and Griffin, A. (2005). *The PDMA Handbook of New Product Development*. doi:10.1002/9780470172483.

Kang, S. W., Sane, C., Vasudevan, N., and Tucker, C. S. (2013). Product resynthesis: Knowledge discovery of the value of end-of-life assemblies and subassemblies. *ASME Journal of Mechanical Design*, *136*(1). doi:10.1115/1.4025526

Karniel, A., & Reich, Y. (2012). Multi-level modelling and simulation of new product development processes. *Journal of Engineering Design*, *24*(3), 185–210.

Kim, K.-Y., Manley, D. G., and Yang, H. (2006). Ontology-based assembly design and information sharing for collaborative product development. *Computer-Aided Design*, *38*(12), 1233–1250.

Kurtoglu, T., Campbell, M. I., Arnold, C. B., Stone, R. B., and McAdams, D. A. (2009). A component taxonomy as a framework for computational design synthesis. *Journal of Computing and Information Science in Engineering*, *9*(1). doi:10.1115/1.3086032

Liang, Y., Tan, R., and Ma, J. (2008). Patent analysis with text mining for TRIZ. In *Proceedings of the 4th IEEE International Conference on Management of Innovation and Technology* (pp. 1147–1151). doi:10.1109/ICMIT.2008.4654531

Martin, J. N. (2000). Processes for engineering a system: An overview of the ANSI/EIA 632 standard and its heritage. *Systems Engineering*, *3*(1). doi:10.1002/(SICI)1520–6858(2000)3:1<1::AID-SYS1>3.0.CO;2-0

McAdams, D. A., Stone, R. B., and Wood, K. L. (1999). Functional interdependence and product similarity based on customer needs. *Research in Engineering Design*, *11*(1), 1–19.

Menon, R., Tong, L. H., Sathiyakeerthi, S., Brombacher, A., and Leong, C. (2003). The needs and benefits of applying textual data mining within the product development process. *Quality and Reliability Engineering International*, *20*(1), 1–15.

Mudambi, S. M., and Schuff, D. (2010). What makes a helpful online review? A study of customer reviews on amazon.com. *MIS Quarterly*, *34*(1), 185–200. Retrieved from http://dl.acm.org/citation.cfm?id=2017447.2017457.

Munková, D., Michal, M., and Martin, V. (2014). Influence of stop-words removal on sequence patterns identification within comparable corpora. In V. Trajkovik and M. Anastas (eds.), *Advances in Intelligent Systems and Computing* (Vol. *231*, pp. 66–76). Springer, Switzerland. doi:10.1007/978-3-319-01466-1_6

Murphy, J., Fu, K., Otto, K., Yang, M., Jensen, D., and Wood, K. (2014). Function based design-by-analogy: A functional vector approach to analogical search. *Journal of Mechanical Design*, *136*(10), 101102.

Nagle, J. G. (1996). *Handbook for Preparing Engineering Documents: From Concept to Completion*. John Wiley.

Paul, M. J., and Dredze, M. (2014). Discovering health topics in social media using topic models. *PloS One*, *9*(8). doi:10.1371/journal.pone.0103408

Petiot, J.-F., and Grognet, S. (2006). Product design: A vectors field-based approach for preference modelling. *Journal of Engineering Design*, *17*(3), 217–233.

Pimmler, T. U. (1994). A development methodology for product decomposition and integration. Research Report, Department of Mechanical Engineering, Massachusetts Institute of Technology, Cambridge, MA.

Pimmler, T. U., and Eppinger, S. D. (1994). Integration analysis of product decompositions. In *Proceedings of the 1994 ASME Design Theory and Methodology Conference (DTM'94)* pp. 342–351). New York: The American Society of Mechanical Engineers.

Ramage, D., and Rosen, E. (2009). Stanford topic modeling toolbox. Retrieved from http://nlp.stanford.edu/software/index.shtml.

Rockwell, J. A., Witherell, P., Grosse, I., and Krishnamurty, S. (2008). A web-based environment for documentation and sharing of engineering design knowledge. In *Proceedings of the ASME Design Engineering Technical Conference and Computers and Information in Engineering Conference* (pp. 671–683). Brooklyn, NY. doi:10.1115/DETC2008-50086

Romanowski, C. J., and Nagi, R. (2004). A data mining approach to forming generic bills of materials in support of variant design activities. *Journal of Computing and Information Science in Engineering*, 4(4), 316–328. doi:10.1115/1.1812556.

Sangelkar, S., and McAdams, D. A. (2013). Mining functional model graphs to find product design heuristics with inclusive design illustration. *Journal of Computing and Information Science in Engineering*, *13*(4). doi:10.1115/1.4025469

Schilling, M. A. (2000). Toward a general modular systems theory and its application to interfirm. *Academy of Management Review*, *25*(2), 312–334.

Sen, C., Summers, J. D., and Mocko, G. M. (2010). Topological information content and expressiveness of function models in mechanical design. *Journal of Computing and Information Science in Engineering*, *10*(3). doi:10.1115/1.3462918

Sharman, D. M., and Yassine, A. A. (2004). Characterizing complex product architectures. *Systems Engineering*, *7*(1), 35–60.

Sheldon, J. G. (2009). *How to Write a Patent Application*. New York: Practising Law Institute.

Sheremetyeva, S., Nirenburg, S., and Nirenburg, I. (1996). Generating patent claims from interactive input. In *Proceedings of the Eighth International Workshop on Natural Language Generation* (pp. 61–70). Philadelphia, PA: The Association for Computational Linguistics (ACL).

Sosa, M. E., Eppinger, S. D., and Rowles, C. M. (2003). Identifying modular and integrative systems and their impact on design team interactions. *Journal of Mechanical Design*, *125*(2), 240.

Sosa, M. E., Eppinger, S. D., and Rowles, C. M. (2004). The misalignment of product architecture and organizational structure in complex product development. *Management Science*, *50*(12), 1674–1689.

Steward, D. V. (1981). Design structure system: A method for managing the design of complex systems. *IEEE Transactions on Engineering Management*, *28*(3), 71–74.

Stone, R. B., and Wood, K. L. (2000). Development of a functional basis for design. *Journal of Mechanical Design*, *122*, 359.

Stone, R. B., Kurtadikar, R., Villanueva, N., and Arnold, C. B. (2008). A customer needs motivated conceptual design methodology for product portfolio planning. *Journal of Engineering Design*, *19*(6), 489–514.

Stone, R. B., Wood, K. L., and Crawford, R. H. (1999). Product architecture development with quantitative functional models. In *Proceedings of the ASME 1999 International Design Engineering Technical Conferences*. New York: The American Society of Mechanical Engineers.

Suh, N. P. (1984). Development of the science base for the manufacturing field through the axiomatic approach. *Robotics & Computer-Integrated Manufacturing*, *1*(3), 397–415.

Taura, T., and Nagai, Y. (2013). Synthesis of functions: Practice of concept generation (3). In *Concept Generation for Design Creativity*. London: Springer.

Thevenot, H., and Simpson, T. (2006). A comprehensive metric for evaluating component commonality in a product family. In *Proceedings of the ASME International Design Engineering Technical Conferences and Computers and Information in Engineering Conference* (pp. 823–832). Philadelphia, PA: ASME.

Toutanova, K., and Manning, C. D. (2000). Enriching the knowledge sources used in a maximum entropy part-of-speech tagger. In *Proceedings of the 2000 Joint SIGDAT Conference on Empirical Methods in* Natural Language Processing and Very Large Corpora: Held in Conjunction with the 38th Annual Meeting of the Association for Computational Linguistics (Vol. *13*, pp. 63–70). doi:10.3115/1117794.1117802

Tseng, Y., Lin, C., and Lin, Y. (2007). Text mining techniques for patent analysis. *Information Processing & Management*, *43*(5), 1216–1247.

Tuarob, S., and Tucker, C. S. (2013). Fad or here to stay: Predicting product market adoption and longevity using large scale, social media data. In *Proceedings of the ASME International Design Engineering Conferences and Computers and Information in Engineering Conference*. Portland, OR. doi:10.1115/DETC2013-12661

Tuarob, S., and Tucker, C. S. (2014). Discovering next generation product innovations by identifying lead user preferences expressed through large scale social media data. In *Proceedings of the ASME International Design Engineering Technical Conferences and Computers and Information in Engineering Conference*. Buffalo, NY. doi:10.1115/DETC2014-34767

Tuarob, S., and Tucker, C. S. (2015a). Automated discovery of lead users and latent product features by mining large scale social media networks. *Journal of Mechanical Design*, *137*(7). doi:10.1115/1.4030049

Tuarob, S., and Tucker, C. S. (2015b). Quantifying product favorability and extracting notable product features using large scale social media data. *Journal of Computing and Information Science in Engineering*, *15*(3). doi:10.1115/1.4029562

Tuarob, S., Tucker, C. S., Salathe, M., and Ram, N. (2013). Discovering health-related knowledge in social media using ensembles of heterogeneous features. In *Proceedings of the 22nd ACM International Conference on Conference on Information and Knowledge Management (CIKM '13)* (pp. 1685–1690). doi:10.1145/2505515.2505629

Tucker, C. S., and Kang, S. W. (2012). A bisociative design framework for knowledge discovery across seemingly unrelated product domains. In *Proceedings of the ASME 2012 International Design Engineering Conferences and Computers and Information in Engineering Conference*. Chicago, IL. doi:10.1115/DETC2012-70764

Tucker, C. S., and Kim, H. M. (2011). Trend mining for predictive product design. *Journal of Mechanical Design*, *133*(11). doi:10.1115/1.4004987

Umeda, Y., Kondoh, S., Shimomura, Y., and Tomiyama, T. (2005). Development of design methodology for upgradable products based on function-behavior-state modeling. *AIE EDAM*, *19*(03), 161–182.

Wei, C.-P., Chen, Y.-M., Yang, C.-S., and Yang, C. C. (2009). Understanding what concerns consumers: A semantic approach to product feature extraction from consumer reviews. *Information Systems and e-Business Management*, *8*(2), 149–167.

Yang, C.-S., Wei, C.-P., and Yang, C. C. (2009). Extracting customer knowledge from online consumer reviews: A collaborative-filtering-based opinion sentence identification approach. In *Proceedings of the 11th International Conference on Electronic Commerce* (pp. 64–71). New York: The American Society of Mechanical Engineers.

Yanhong, L., and Runhua, T. (2007). A text-mining-based patent analysis in product innovative process. In N. León-Rovira and S. K. Cho (eds.), *Trends in Computer Aided Innovation* (Vol. 250, pp. 89–96). New York: Springer.

Yassine, A. A., and Braha, D. (2003). Complex concurrent engineering and the design structure matrix method. *Concurrent Engineering*, *11*(3), 165–176.

Yoon, B., and Park, Y. (2004). A text-mining-based patent network: Analytical tool for high-technology trend. *Journal of High Technology Management Research*, *15*(1), 37–50.

Zhan, J., Loh, H. T., and Liu, Y. (2009). Gather customer concerns from online product reviews: A text summarization approach. *Expert Systems with Applications*, *36*(2), 2107–2115.

Zhang, X. (Luke), Simpson, T. W., Frecker, M., and Lesieutre, G. (2012). Supporting knowledge exploration and discovery in multi-dimensional data with interactive multiscale visualisation. *Journal of Engineering Design*, *23*(1), 23–47.

Zhou, F., Jianxin Jiao, R., and Linsey, J. S. (2015). Latent customer needs elicitation by use case analogical reasoning from sentiment analysis of online product reviews. *Journal of Mechanical Design*, *137*(7). doi:10.1115/1.4030159.

section five

Work integration

chapter nine

Theoretical framework for work integration

This chapter presents a theoretical framework to approach the integration of cognitive work design in manufacturing. The framework is a systematic and iterative approach to evaluating and finding a balance for the cognitive load on the operator to meet work demands.

Cognitive evaluation in manufacturing

Manufacturing is a human-driven transformation process that uses energy and manpower to produce consumer goods, which includes the tasks of production, assembly, logistics, planning, maintenance, and quality management (Spath et al., 2012). Today's manufacturing and assembly processes must be flexible in order to adapt quickly to a growing number of customized product types and changing market demands (Bannat et al., 2011). Flexible work entails multitasking and rapid changes in work conditions (Hoc, 2008).

Due to the growing demand for flexible production systems, adaptive interfaces for the optimal support of production workers in the manufacturing environment have become increasingly relevant (Stork et al., 2007). Therefore, when developing a system to support the operator, it is important that the system provides sufficient information at the correct time so that the operator can receive and process it with minimal effort. For example, in the manual assembly process, the system must be aware of the environment and the current state of the product in the manufacturing process in addition to incorporating data on the cognitive processes involved during manual assembly (Stoessel et al., 2008). Evaluating the cognitive processes ensures that the support systems do not overload the operator. In these complex production environments, operators are required to filter multiple sources of information, attentively decide on relevant information, incorporate perceptual information with action goals, monitor these tasks in their working memory, and control their appropriate response actions (Stork et al., 2007).

In fast-paced manufacturing operations, there is the potential for operator errors, which are associated with possible safety issues and lost revenue. Manufacturing operations are moving away from force-focused physical activity to cognitive control activity (Spath et al., 2012). Cognitive control activity applies cognitive ergonomics, which studies work processes with a focus on understanding a situation in order to support reliable, effective, and satisfactory performance (Martin et al., 2011). It assesses problems relative to attention distribution, decision making, the formation of learning skills, the usability of human–computer systems, the cognitive aspects of mental load, stress, and human errors during work (Martin et al., 2011).

Multitasking can be defined as the result of time allocation decisions that humans make when they are confronted with more than one task (Benbunan-Fich et al., 2011). It is a common activity in human–machine interaction; control room operators in the manufacturing industry may operate a device and monitor several displays at the same time (Wu and Liu, 2009). Since multitasking has become more prevalent in manufacturing operations, the need for cognitive work analysis must be examined in human–system design. Also, with the need for increasing production speeds to maintain global competitiveness, the cognitive limits of human performance must be considered in work design (Allwood et al., 2016). This is an important consideration, because both mental overload and underload are associated with performance degradation, and optimal work design will keep the workload within performance range, where the workload is neither too high nor too low (Stanton et al., 2004). This association of optimal work design for improved performance coincides with the findings originally developed by Robert M. Yerkes and John D. Dodson, in 1908, to measure performance and arousal levels, which affect stress, anxiety, and motivation. The Yerkes–Dodson law indicates that improved performance can result from arousal, as long as the arousal levels are not too low or too high (Teigen, 1994), which can lead to detriment in decision making as well as performance.

The evaluation of cognitive ergonomics in manufacturing operations is limited in the literature. However, this topic is pertinent for modern work design because, according to Allwood et al. (2016, p. 750):

> Despite the extraordinary adaptability of human beings, the speed of their cognitive processes has limits, and in turn this will provide a limit to the absolute speed at which manufacturing can occur: despite extensive "hype" about robots taking over human jobs, the reality of the past 40 years of automation is that robots or other computer operated systems are less able than humans to respond to unfamiliar situations, so while highly controlled and repeatable tasks can be automated, it is unlikely that manufacturing will ever be independent of

human control, support, innovation, leadership and repair.

The implication of cognitive ergonomics in manufacturing is that it can improve the efficiency of the human–machine system; therefore, work design should incorporate cognitive ergonomic techniques in an effort to maximize performance.

Performance optimization influences safety in the work environment and profitability. Table 9.1 presents a few studies that have previously investigated the topic of cognitive ergonomics in manufacturing.

Although there is a large body of literature on physical ergonomics for process improvements, the findings on cognitive ergonomics relative to mental workload (MWL) in manufacturing are limited. This chapter will present a systematic approach for measuring MWL in manufacturing, using a combination of analytical and empirical techniques. This approach uses mathematical modeling, along with physiological, subjective, and performance measures, to evaluate mental load. Mathematical

Table 9.1 Cognitive ergonomics studies in manufacturing

Authors	Year of Publication	Methods
Bommer and Fendley	2015	Conducted a pilot study to test a framework design for MWL resource evaluation in manufacturing processes
Lindblom and Thorvald	2014	Developed a framework to evaluate cognitive work environment problems (CWEPs) that affect cognitive load in manufacturing
Thorvald and Lindblom	2014	Development of a cognitive load assessment tool to identify risks of tasks in a workstation design
Tan et al.	2009	Conducted experiments to investigate MWL in human–robot collaboration (HRC) for cellular manufacturing
Layer et al.	2009	Developed a predictive algorithm to simulate the human performance of an individual (groups of individuals) relative to cognitive demands and the quality of work life attributes
Stoessel et al.	2008	Conducted a study to test a cognitive assistance system for improving performance in manual assembly tasks
Genaidy and Karwowski	2003	Developed a framework to guide future research on the impact of lean production strategies on work demands

modeling is used to provide early predictions of MWL for evaluating and improving the human–system design for a shop floor employee. This systematic approach presents a framework that evaluates the effect of cognitive resources on human performance in manufacturing operations applying multiple resource theory (MRT) and MWL measures.

Theoretical framework

The Mental Resource Assessment in Manufacturing (M-RAM) framework, presented in Figure 9.1, is made up of six primary steps (Bommer, 2016). In step 1, the manufacturing process is examined to understand the work system and its task elements. Any of the cognitive task analysis methods discussed in Chapter 5 can be used as tools for this evaluation, as well as hierarchical task analysis (HTA), discussed in Chapter 7. Next, MRT is employed in step 2 to determine the mental resource utilization required to accomplish the work elements; the MRT scale in the Improved Performance Research Integration Tool (IMPRINT) is applied to this evaluation. Mathematical modeling is deployed in step 3 to establish the expected MWL in the work system design. As discussed in Chapter 7, there are a number of modeling tools that can be used. However, this framework uses IMPRINT because of its underpinning theory of MRT. Step 4 applies human-in-the-loop (HITL) simulations to measure MWL using subjective, performance, and physiological measures (various measures were discussed in Chapter 6). The engineer or analyst can apply any methods from those categories that best fit the work system under analysis. During step 5, the data from steps 3 and 4 are analyzed to determine whether the process is overloaded or underloaded. If there is an MWL overload, an analysis using MRT should be performed to establish the attribution of mental resources that are associated with the mental overload and/or performance errors. In the event that a mental overload condition is not found, the process is finished. However, if there is a resulting mental overload condition, the manufacturing system can be redesigned (step 6) using MRT principles to reduce the resource utilization associated with the mental overload, and the MRT rating scale will be applied again to compare the redesigned system resources with the baseline. Similarly, for capacity planning, if an underloaded mental condition occurs, the process can be modified and mapped back to MRT. Once more, the MWL is measured. This loop carries on until an optimal workload range is reached. Mitchell (2000) describes optimal workload as "a situation in which the operator feels comfortable, can manage task demands intelligently, and maintain good performance" (Hart, 1991, p. 3).

As manufacturing operations move in the direction of more complex systems and procedures to become more competitive and adaptive to production demands, cognitive demands are becoming an integral component

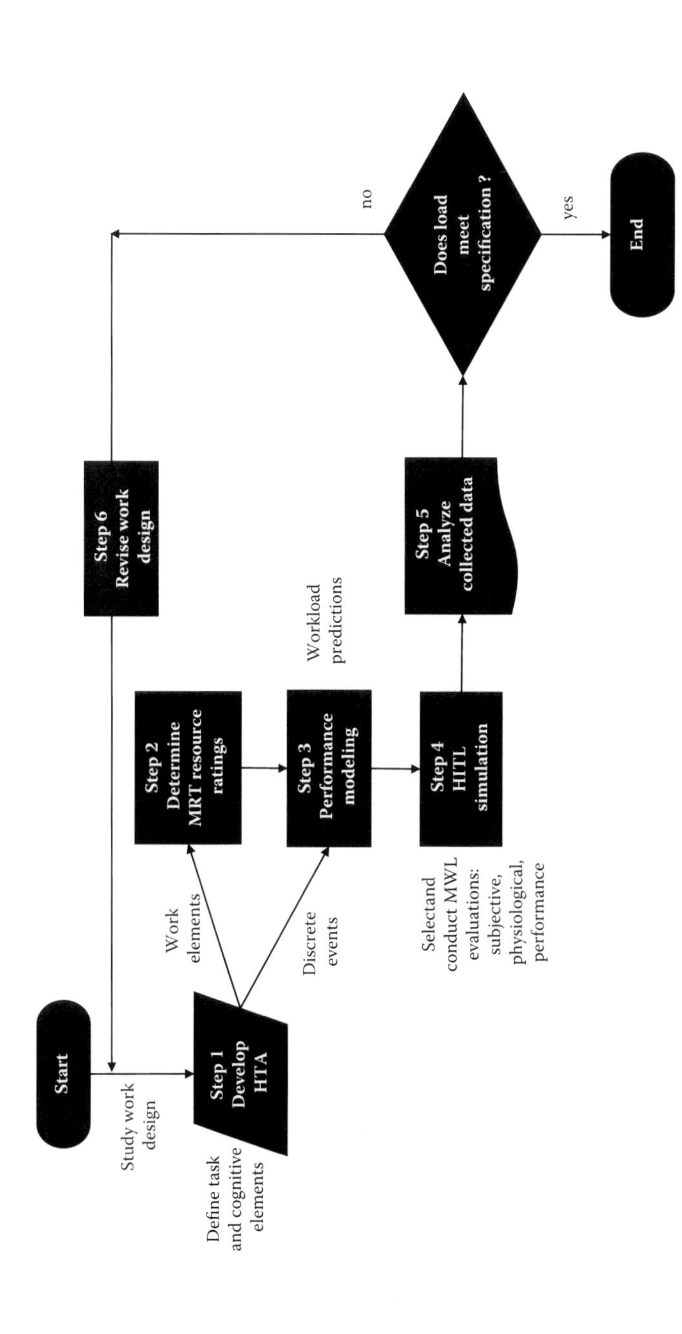

Figure 9.1 M-RAM procedural steps to the framework. (Adapted from S. C. Bommer, A theoretical framework for evaluating mental workload resources in human systems design for manufacturing operations, 2016.)

for manufacturing operations. Therefore, cognitive models are helpful for analyzing the operator's performance as it relates to MWL while performing multitasking functions. Theoretical frameworks to assist the development of these cognitive models are needed to offer a better understanding of the human–system interaction that affects an operator's MWL.

Framework validation

Pilot study

A pilot study was conducted in the field to assess and support the need for cognitive-based design in manufacturing. The pilot is described in detail by Bommer (2016). The framework applied a systematic approach to evaluating cognitive load in the medical device domain for the production of a medical surgical implant for humans. After the devices were laser welded, a postprocessing procedure was required to clean and check the quality specifications of the parts. This process was repetitive, such that the same procedure had to be followed each cycle. This study analyzed a segment of the process: parts cleaning. Process steps for parts cleaning include mixing acids for proper formulation, dipping parts in chemicals, and monitoring equipment settings to track proper processing times. To study the manufacturing system, experienced real-world process operators who regularly performed the part cleaning process were utilized in on-site interviews and process walkthroughs for data collection to test the initial form of the theoretical framework.

Figure 9.2 lists the five primary steps of the systematic approach of the theoretical framework used in the pilot study. In step 1, the manufacturing process was investigated to acquire information on the work system. An HTA was completed to understand the task elements that were being assessed. Next, an applied cognitive task analysis (ACTA) was utilized to evaluate the cognitive elements in the human system design. The HTA and ACTA were used jointly to define the process steps and cognitive elements, which were the discrete events in IMPRINT. During step 2, MRT was applied using the scale in IMPRINT to assess the mental resources

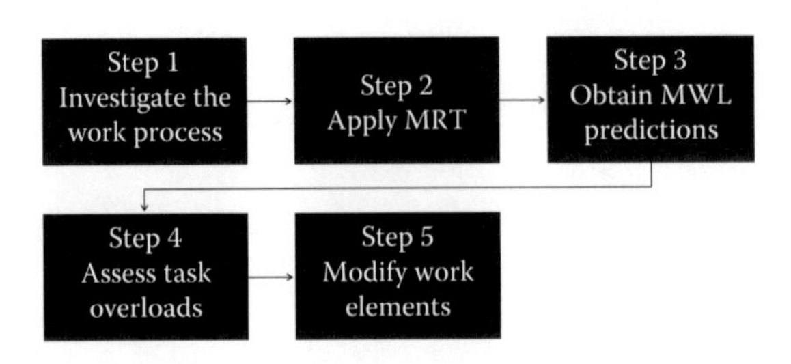

Figure 9.2 Procedural steps of the pilot study.

required for completion of the process task's discrete events. This step assigns MRT ratings for each discrete event itemized in step 1; these ratings were the inputs for IMPRINT. Next, in step 3, MWL predictions were obtained by simulating the process in IMPRINT. The discrete events, with their estimated task times, and the MRT ratings for the associated mental resources were the primary simulation inputs. With these data, IMPRINT provided the workload predictions and a workload profile for the defined task. To mitigate a MWL overload condition, an analysis using MRT was done in step 4 to verify the attribution of mental resources that correlate with the mental overload. If a mental overload condition was not measured, this would terminate the process. In the event of a mental overload situation, the manufacturing system would be modified, during step 5, based on MRT principles, to reduce the resources contributing to the mental overload; MRT would then be applied again to compare the modified system resources with the baseline. Once again, MWL is measured using IMPRINT. This loop continues until an optimal solution range is achieved. Once an optimal range is attained, a different variation of work elements is tested on the shop floor. The different work element variations and their associated load could be documented to form a cognitive ergonomic index for a quick design reference.

The results of the pilot study predicted that overlapping more than two inspection tasks during the cleaning process would be unmanageable by the operator, because a three-task overlap created a workload that surpassed the threshold. This means the predicted workload value of a three-task overlap for inspection could potentially put the operator in a state of high load, which could potentially create inefficiencies in the process. These findings support the idea that the proposed framework could be applied during the design and setup phases of a process to construct better operations for evaluating the mental overload of the operator in repetitive task processes. When evaluating the results of this pilot study, this model can be valuable for understanding operator MWL resources in human–machine design. This is useful for the capacity planning of process setups in systems that have yet to be built and for modifications to existing systems, to improve the design of the modern manufacturing processes that necessitate more cognitive demands. Also, more studies should be performed to validate the appropriate MWL threshold by analyzing where increased task demands degrade performance.

In the next section, the framework was expanded, as depicted in the preceding section "Theoretical Framework," by adding performance measures and workload assessments for further validation. Subjective workload assessment techniques and physiological measures will be used to verify the mathematical simulation results. In order to gather physiological data to support the subjective workload measures, eye tracking will be exploited in the next study.

Laboratory experiment

Bommer (2016) conducted a within-subjects design experiment to validate the M-RAM framework, using a repetitive task simulation composed of toy building blocks in a laboratory setting at Wright State University in the Human Performance and Cognition Laboratory. The experiment was designed and implemented using the process steps and flow of the M-RAM framework described in "Theoretical Framework." The procedural steps of the experiment are outlined in Figure 9.3.

Prior to beginning the experiment, each subject was provided with a briefing on the test scenarios and procedures. Next, the protocol for informed consent was followed. For each simulated scenario (i.e., treatment) the subjects were allowed up to 5 minutes to review the work instructions and train with the toy building blocks and inspection tools. Toy building blocks were chosen as the experimental apparatus because of their similarity to an assembly process. During the HITL simulation, the subjects had to follow a combination of color criteria and use a scale and caliper to obtain specified inspection measurements. There were different instructions for each test scenario. The color coding and tool measurements were used to simulate inspection tasks. Next, eye-tracking glasses were used to collect fixation frequency measures. The eye-tracking system was calibrated before collecting data from each treatment. Immediately following each system calibration, an experimental treatment was performed. Each treatment was given a targeted cycle time, although the process was self-paced. Each cycle was completed five consecutive times during each treatment. After completing each scenario, the analyst provided the subject with a NASA-TLX questionnaire to complete. Table 9.2 outlines the dependent and independent variables in the experiment.

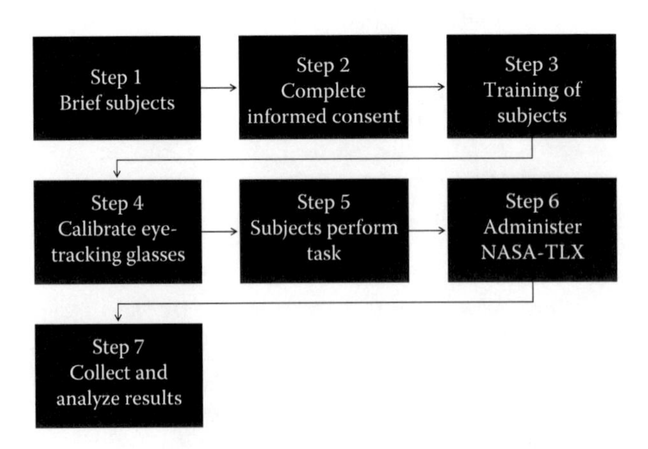

Figure 9.3 Procedural steps of the laboratory experiment.

Table 9.2 Experimental variables

Type	Variable	Measure
Independent	Task complexity	Assembly and inspection with tools
		Assembly with tools
		Assembly and inspection
		Assembly only
Dependent	Mental workload	Subjective: NASA-TLX
		Physiological: fixation frequency
		Performance: human error probability (HEP)

The independent variable of task complexity was made up of four levels. The levels consisted of various combinations and quantities of work elements, including inspection and assembly tasks. The dependent variable of MWL consisted of a measure from the three categories of MWL discussed in Chapter 6 (subjective, physiological, and performance). Each of the response variables for this experiment is an element in the theoretical framework. The process flow of the theoretical framework (Figure 9.1) was applied to develop and examine the different conditions of the simulated processes in this experiment. A statistical software package was used to analyze the response variables. The experimental treatments were randomized to minimize any nuisance variables.

In summary, this experiment simulated a manual repetitive manufacturing process with inspection tasks in a laboratory setting to validate the M-RAM framework. The framework was validated using mathematical modeling and MWL measures. As part of the validation, four experimental scenarios (i.e., treatments) with toy building blocks were used, which varied in difficulty by the type and number of work elements imposed on the operator. The key finding of the experiment is that the MWL inputs of the M-RAM framework all predicted a statistically significant difference between the different treatment levels for each task modeled. A correlation analysis was conducted to evaluate linearity among the MWL measures, and two of the three measures were highly correlated, although all three measures showed significant differences across treatment levels. Therefore, when applying the framework, it is important to use all the measurement categories to make MWL predictions. The results of the experimental findings support the hypothesis that the M-RAM framework accurately predicts MWL for repetitive manufacturing tasks.

In addition, the findings of this study highlighted the influence of inspection tasks on the mental load of an operator during a manual assembly procedure. Although the simulated task would be considered rather simple, when the inspection tasks were inserted into the work design, the cognitive load of the operator peaked. An extensive study applying this framework (Figure 9.1) could benefit the manufacturing industry by developing regression models for practical applications to optimize cognitive load in work design, which could improve safety and performance.

Summary

The objective of this M-RAM framework is to present the theory of cognitive ergonomics and its applications to the evaluation of operator MWL resources in manufacturing operations, which involve repetitive tasks and require multitasking, in an effort to improve operator performance and human–system design. This approach provides an alternative method for system design and work development, as multitasking is becoming an important function of production processes. Work system design that requires simple routine procedures must be approached in a different manner from tasks with higher cognitive load, in order to maintain the desired levels of performance. Eklund (2000) demonstrates a connection between workers' job characteristics, which include physical, physiological, cognitive, and other organizational factors, and its associated quality and productivity performance measures.

The study of cognitive load in manufacturing operations is a relatively new focus in cognitive ergonomics, and the literature is limited in regard to this type of cognitive evaluation. Therefore, this framework is vital for providing a method, which uses MRT as the underpinning theory, of evaluating and mitigating cognitive load in repetitive task manufacturing operations that require multitasking. It is an innovative approach in manufacturing for human systems design, because this framework integrates cognitive load into the manufacturing work design. It is applicable to the changing demands of the manufacturing industry and can be utilized throughout various domains. Bommer (2016) first introduced this framework during a pilot study. Since that time, the theoretical framework has been further developed and validated in laboratory settings, which justifies future work with more extensive studies and validation in the field. In closing, the concept of cognitive ergonomics finds a balance between the human's cognitive abilities and limitations, as well as the machine, task, and environment (Kramer, 2009). Therefore, the presented framework is a tool for finding this balance in work design.

References

Allwood, J. M., Childs, T. H., Clare, A. T., De Silva, A. K., Dhokia, V., Hutchings, I. M., et al. (2016). Manufacturing at double the speed. *Journal of Materials Processing Technology*, 229, 729–757.

Bannat, A., Bautze, T., Beetz, M., Blume, J., Diepold, K., Ertelt, C., et al. (2011). Artificial cognition in production systems. *IEEE Transactions on Automation Science and Engineering*, 8(1), 148–174.

Benbunan-Fich, R., Adler, R. F., and Mavlanova, T. (2011). Measuring multi-tasking behavior with activity-based metrics. *ACM Transactions on Computer–Human Interaction (TOCHI)*, 18(2), 7.

Bommer, S. C. (2016). A theoretical framework for evaluating mental workload resources in human systems design for manufacturing operations. Electronic thesis or dissertation. Retrieved from https://etd.ohiolink.edu.

Bommer, S. C., and Fendley, M. (2015). Reducing mental workload resources in human systems design for manufacturing operations. *Proceedings of the IIE Annual Conference* 2015, pp. 1654.

Eklund, J. (2000). Development work for quality and ergonomics. *Applied Ergonomics*, 31(6), 641–648.

Genaidy, A. M., and Karwowski, W. (2003). Human performance in lean production environment: Critical assessment and research framework. *Human Factors and Ergonomics in Manufacturing & Service Industries*, 13(4), 317–330.

Hart, S. G. (1991). Pilots' workload coping strategies. In AIAA/NASA/FAA/HFS Conference on Challenges in Aviation Human Factors: The National Plan.

Hoc, J. (2008). Cognitive ergonomics: A multidisciplinary venture. *Ergonomics*, 51(1), 71–75.

Kramer, A. (2009). An introduction to cognitive ergonomics. Ask Ergo Works. Retrieved from http://old.askergoworks.com/news/18/Cognitive-Ergonomics-101-Improving-Mental-Performance.aspx.

Layer, J. K., Karwowski, W., and Furr, A. (2009). The effect of cognitive demands and perceived quality of work life on human performance in manufacturing environments. *International Journal of Industrial Ergonomics*, 39(2), 413–421.

Lindblom, J., and Thorvald, P. (2014). Towards a framework for reducing cognitive load in manufacturing personnel. *Advances in Cognitive Engineering and Neuroergonomics*, 11, 233–244.

Martin, P. R., Cheung, F. M., Knowles, M. C., Kyrios, M., Littlefield, L., and Overmier, J. B. (2011). *IAAP Handbook of Applied Psychology*. Chichester, UK: John Wiley.

Mitchell, D. K. (2000). Mental workload and ARL workload modeling tools (Technical Note ARL-TN-161). Aberdeen Proving Ground, MD: US Army Research Laboratory.

Spath, D., Braun, M., and Meinken, K. (2012). Human factors in manufacturing. *Handbook of Human Factors and Ergonomics* (4th ed.), pp. 1643–1666.

Stanton, N. A., Hedge, A., Brookhuis, K., Salas, E., and Hendrick, H. W. (2004). *Handbook of Human Factors and Ergonomics Methods*. Boca Raton, FL: CRC Press.

Stoessel, C., Wiesbeck, M., Stork, S., Zaeh, M. F., and Schuboe, A. (2008). Towards optimal worker assistance: Investigating cognitive processes in manual assembly. In *Manufacturing Systems and Technologies for the New Frontier*, pp. 245–250. London: Springer.

Stork, S., Stobel, C., Muller, H., Wiesbeck, M., Zah, M., and Schubo, A. (2007). A neuroergonomic approach for the investigation of cognitive processes in interactive assembly environments. *Proceedings of the 16th IEEE International Symposium on Robot and Human Interactive Communication (RO-MAN)*, 2007, pp. 750–755.

Tan, J. T. C., Duan, F., Zhang, Y., Watanabe, K., Kato, R., and Arai, T. (2009). Human–robot collaboration in cellular manufacturing: Design and development. *Proceedings of the 2009 IEEE/RSJ International Conference on Intelligent Robots and Systems*, pp. 29–34.

Teigen, K. H. (1994). Yerkes–Dodson: A law for all seasons. *Theory & Psychology*, 4(4), 525–547.

Thorvald, P., and Lindblom, J. (2014). Initial development of a cognitive load assessment tool. *Proceedings of the 5th AHFE International Conference on Applied Human Factors and Ergonomics, 19–23 July 2014, Krakow, Poland*, pp. 223–232.

Wu, C., and Liu, Y. (2009). Development and evaluation of an ergonomic software package for predicting multiple-task human performance and mental workload in human–machine interface design and evaluation. *Computers & Industrial Engineering*, 56(1), 323–333.

chapter ten

Project management for work management

Every work element should be managed as a project. Project management is work management and vice versa. If they are not managed like projects, some work elements run the risk of being means to no end. Badiru (2016) comments, "It is our natural biological imperative to work [...] unfortunately there is no imperative for that work to be useful." Project management is the process of managing, allocating, and timing resources in order to achieve a given objective in an expedient manner. The objective may be stated in terms of time (schedule), performance output (quality), or cost (budget). It is the process of achieving objectives by utilizing the combined capabilities of available resources. Time is often the most critical aspect of managing any project. Time is the physical platform on which project accomplishments are made. So, it must be managed concurrently with any other important aspects in a project. Project management covers the following basic functions:

1. Planning
2. Organizing
3. Scheduling
4. Control

The complexity of a project can range from simple, such as the painting of a vacant room, to very complex, such as the introduction of a new high-tech product. The technical differences between project types are of great importance when selecting and applying project management techniques. Figure 10.1 illustrates the various dimensions of the application of project management to an industrial system.

Project management techniques are widely used in many human endeavors, such as construction, banking, manufacturing, marketing, health care, sales, transportation, and research and development, as well as in academic, legal, political, and government establishments, just to name a few. In many situations, the on-time completion of a project is of paramount importance. Delayed or unsuccessful projects not only translate to monetary losses but also impede subsequent undertakings. Project

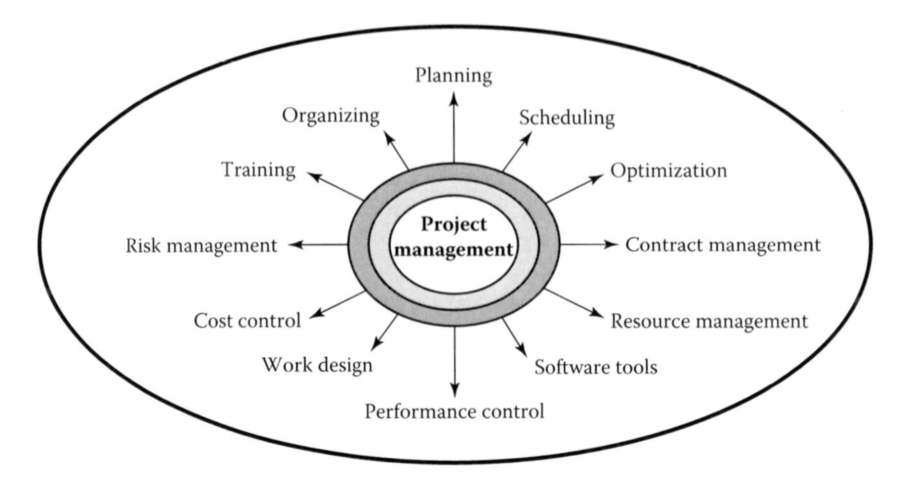

Figure 10.1 Multiple dimensions of project management.

management takes a hierarchical view of a project environment, covering the following top-down levels:

1. System level
2. Program level
3. Project level
4. Task level
5. Activity level

Project review and selection criteria

Project selection is an essential first step in focusing the efforts of an organization. Figure 10.2 presents a simple graphical evaluation of project selection. The vertical axis represents the value-added basis of the project under consideration, while the horizontal axis represents the level of complexity associated with the project. In this example, value can range from low to high, while complexity can range from easy to difficult. The figure shows four quadrants containing regions of high value with high complexity, low value with high complexity, high value with low complexity, and low value with low complexity. A fuzzy region is identified with an overlay circle. The organization must evaluate each project on the basis of overall organization value streams. The figure can be modified to represent other factors of interest to an organization instead of value-added and project complexity.

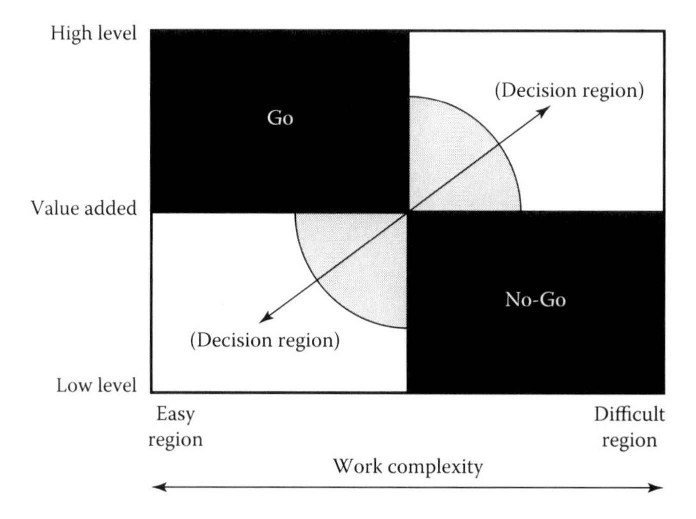

Figure 10.2 Project selection quadrant: *go* or *no-go*.

Criteria for project review

Some of the specific criteria that may be included in project review and selection are as follows:

- Cost reduction
- Customer satisfaction
- Process improvement
- Revenue growth
- Operational responsiveness
- Resource utilization
- Project duration
- Execution complexity
- Cross-functional efficiency
- Partnering potential

Hierarchy of work selection

In addition to evaluating an overall project, the elements making up the project may need to be evaluated on the basis of the following hierarchy. This will facilitate achieving an integrated project management view of the organization's operations.

- System
- Program

- Task
- Work packages
- Activity

Sizing of projects

Associating a size measure with an industrial project provides a means of determining the level of relevance and the effort required. A simple guideline is as follows:

- Major (over 60 man-months of effort)
- Intermediate (between 6 and 60 man-months)
- Minor (less than 6 man-months)

Planning levels

When selecting projects and their associated work packages, planning should be done in an integrative and hierarchical manner using the following levels of planning:

- Supra level
- Macro level
- Micro level

Hammersmith's project alert scale: Red, yellow, green convention

Hammersmith (2006) presented a guideline alert scale for project tracking and evaluation. He suggested putting projects into categories of *red*, *yellow*, or *green* with the following definitions:

Red: If not corrected, the project will be late and/or over budget.
Yellow: The project is at risk of turning *red*.
Green: The project is on time and on budget.

Product assurance concept for corporate projects

Product assurance activities will provide the product deliverables throughout a program development period. These specific activities for continuous effort are to

1. *Track and incorporate specific technologies*: The technology management task will track pertinent technologies through various means (e.g., vendor surveys and literature research). More importantly, the task will determine strategies to incorporate specific technologies.

2. *Analyze technology trends and conduct long-range planning*: The output of technology assessments should be used to formulate long-range policies, directions, and research activities so as to promote longevity and evolution.
3. *Encourage government and industry leaders' participation*: In order to determine a long-term strategy, technical evaluators need to work closely with government and industry leaders so as to understand their long-range plans. Technology panels may be formed to encourage participation from these leaders.
4. *Influence industry directions*: As with other development programs, program management can be used to influence industry directions and spawn new technologies. Since the effort can be treated as a model, many technologies and products that have been developed can be applied to other similar systems.
5. *Conduct prototyping work*: Prototyping will be used to evaluate the suitability, feasibility, and cost of incorporating a particular technology. In essence, it provides a less costly mechanism to test a technology before significant investment is made in the product development process. Technologies that are high risk with high payoffs should be chosen as the primary subjects for prototyping.

Body-of-knowledge methodology

The Project Management Body of Knowledge (PMBOK®) is published and disseminated by the Project Management Institute (PMI). The body of knowledge comprises specific knowledge areas, which are organized into the following broad areas:

1. Project *integration* management
2. Project *scope* management
3. Project *time* management
4. Project *cost* management
5. Project *quality* management
6. Project *human* resource management
7. Project *communications* management
8. Project *risk* management
9. Project *procurement* management

These segments of the PMBOK cover the range of functions associated with any project, particularly complex ones. Multinational projects, particularly, pose unique challenges pertaining to reliable power supplies, efficient communication systems, credible government support, dependable procurement processes, the consistent availability of technology, progressive industrial climates, trustworthy risk mitigation infrastructures, regular supplies of skilled labor, uniform focus on the quality of work, global consciousness, hassle-free bureaucratic processes, coherent safety and security systems, steady law and order, unflinching focus on customer satisfaction, and fair labor relations. Assessing and resolving concerns about these issues in a step-by-step fashion will create a foundation of success for a large project. While no system can be perfect and satisfactory in all aspects, a tolerable trade-off on these factors is essential for project success.

Components of knowledge areas

The key components of each element of the body of knowledge are as follows:

- Integration
    - Integrative project charters
    - Project scope statements
    - Project management plans
    - Project execution management
    - Change control
- Scope management
    - Focused scope statements
    - Benefit–cost analysis
    - Project constraints
    - Work breakdown structures
    - Responsibility breakdown structures
    - Change control
- Time management
    - Schedule planning and control
    - Program evaluation and review technique (PERT) and Gantt charts
    - Critical path methods
    - Network models
    - Resource loading
    - Reporting
- Cost management
    - Financial analysis
    - Cost estimating
    - Forecasting
    - Cost control
    - Cost reporting
- Quality management
    - Total quality management
    - Quality assurance

    - Quality control
    - Cost of quality
    - Quality conformance
- Human resources management
    - Leadership skill development
    - Team building
    - Motivation
    - Conflict management
    - Compensation
    - Organizational structures
- Communications
    - Communication matrices
    - Communication vehicles
    - Listening and presenting skills
    - Communication barriers and facilitators
- Risk management
    - Risk identification
    - Risk analysis
    - Risk mitigation
    - Contingency planning
- Procurement and subcontracts
    - Material selection
    - Vendor prequalification
    - Contract types
    - Contract risk assessment
    - Contract negotiation
    - Contract change orders

Step-by-step implementation

The efficacy of the systems approach to project management is based on step-by-step and component-by-component implementation of the project management process. The major knowledge areas of project management are administered in a structured outline covering the following six basic clusters: (1) initiating, (2) planning, (3) executing, (4) monitoring, (5) controlling, and (6) closing. The implementation clusters represent five process groups that are followed throughout the project life cycle. Each cluster itself consists of several functions and operational steps. When the clusters are overlaid on the nine knowledge areas, we obtain a two-dimensional matrix that spans 44 major process steps. Table 10.1 shows an overlay of the project management knowledge areas and the implementation clusters. The monitoring and controlling clusters are usually administered as one process group (monitoring and controlling). In some cases, it may be helpful to separate them to highlight the essential attributes of

Table 10.1 Overlay of project management areas and implementation clusters

Knowledge areas	Project management process clusters				
	Initiating	Planning	Executing	Monitoring and controlling	Closing
Project integration	Developing the project charter Developing the preliminary project scope	Developing a project management plan	Directing and managing project execution	Monitoring and controlling project work Integrated change control	
Scope		Scope planning Scope definition Creating a WBS		Scope verification Scope control	
Time		Activity definition Activity sequencing Activity resource estimating Activity duration estimating Schedule development		Schedule control	
Cost		Cost estimating Cost budgeting		Cost control	
Quality		Quality planning	Performing quality assurance	Performing quality control	
Human resources		Human resource planning	Acquiring the project team Developing the project team	Managing project team	
Communication		Communication planning	Information distribution	Performance reporting Stakeholder management	
Risk		Risk management planning Risk identification Qualitative risk analysis Quantitative risk analysis Risk response planning		Risk monitoring and control	
Procurement		Planning purchases and acquisitions Planning contracting	Requesting seller responses Selecting sellers	Contract administration	Contract closure

each cluster of functions over the project life cycle. In practice, the processes and clusters do overlap. Thus, there is no crisp demarcation of when and where one process ends and where another one begins over the project life cycle. In general, the project life cycle defines the following:

1. Resources that will be needed in each phase of the project life cycle
2. Specific work to be accomplished in each phase of the project life cycle

It should be noted that *project* life cycle is distinguished from *product* life cycle. The project life cycle does not explicitly address operational issues, whereas the product life cycle mostly concerns operational issues from the product's delivery to the end of its useful life. Note that for science, technology, and engineering (STE) projects, the shape of the life cycle curve may be expedited due to the rapid developments that often occur in STE activities. For example, for a high-tech project, the entire life cycle may be shortened, with a very rapid initial phase, even though the conceptualization stage may be very long. Typical characteristics of a project life cycle include the following:

1. Cost and staffing requirements are lowest at the beginning of the project and ramp-up during the initial and development stages.
2. The probability of successfully completing the project is lowest at the beginning and highest at the end. This is because many unknowns (risks and uncertainties) exist at the beginning of the project. As the project nears its end, there are fewer opportunities for risks and uncertainties.
3. The risks to the project organization (project owner) are lowest at the beginning and highest at the end. This is because not much investment has gone into the project at the beginning, whereas much has been committed by the end of the project. There is a higher sunk cost manifested at the end of the project.
4. The ability of the stakeholders to influence the final project outcome (cost, quality, and schedule) is highest at the beginning and becomes progressively lower toward the end of the project. This is intuitive because influence is best exerted at the beginning of an endeavor.
5. The value of scope changes decreases over time during the project life cycle, while the cost of scope changes increases over time. The suggestion is to decide and finalize scope as early as possible. If there are to be scope changes, make them as early as possible.

Project systems structure

The overall execution of a project management system is outlined in the following sections.

Problem identification

Problem identification is the stage where a need for a proposed project is identified, defined, and justified. A project may be concerned with the development of new products, the implementation of new processes, or the improvement of existing facilities.

Project definition

Project definition is the phase at which the purpose of the project is clarified. A *mission statement* is the major output of this stage. For example, a prevailing low level of productivity may indicate a need for a new manufacturing technology. In general, the definition should specify how project management may be used to avoid missed deadlines, poor scheduling, inadequate resource allocation, lack of coordination, poor quality, and conflicting priorities.

Project planning

A plan represents the outline of the series of actions needed to accomplish a goal. Project planning determines how to initiate a project and execute its objectives. It may be a simple statement of a project goal or it may be a detailed account of procedures to be followed during the project. Planning can be summarized as

- Objectives
- Project definition
- Team organization
- Performance criteria (time, cost, quality)

Project organization

Project organization specifies how to integrate the functions of the personnel involved in a project. Organizing is usually done concurrently with project planning. Directing is an important aspect of project organization. Directing involves guiding and supervising the project personnel. It is a crucial aspect of the management function. Directing requires skillful managers who can interact with subordinates effectively through good communication and motivation techniques. A good project manager will facilitate project success by directing his or her staff, through proper task assignments, toward the project goal.

Workers perform better when there are clearly defined expectations. They need to know how their job functions contribute to the overall goals of the project. Workers should be given some flexibility for self-direction in performing their functions. Individual worker needs and limitations should be recognized by the manager when directing project functions. Directing a project requires skills dealing with motivating, supervising, and delegating.

Resource allocation

Project goals and objectives are accomplished by allocating resources to functional requirements. Resources can consist of money, people, equipment, tools, facilities, information, skills, and so on. These are usually in short supply. The people needed for a particular task may be committed to other ongoing projects. A crucial piece of equipment may be under the control of another team.

Project scheduling

Timeliness is the essence of project management, and scheduling is often the major focus. The main purpose of scheduling is to allocate resources so that the overall project objectives are achieved within a reasonable time span. Project objectives are generally conflicting in nature. For example, the minimization of the project completion time and the minimization of the project costs are conflicting objectives. That is, one objective is improved at the expense of worsening the other. Therefore, project scheduling is a multiple-objective decision-making problem.

In general, scheduling involves the assignment of time periods to specific tasks within the work schedule. Resource availability, time limitations, urgency levels, required performance levels, precedence requirements, work priorities, technical constraints, and other factors complicate the scheduling process. Thus, the assignment of a time slot to a task does not necessarily ensure that the task will be performed satisfactorily in accordance with the schedule. Consequently, careful control must be developed and maintained throughout the project-scheduling process.

Project tracking and reporting

This phase involves checking whether or not project results conform to project plans and performance specifications. Tracking and reporting are prerequisites for project control. A properly organized report of the project status will help identify any deficiencies in the progress of the project and help pinpoint corrective actions.

Project control

Project control requires that appropriate actions be taken to correct unacceptable deviations from expected performance. Control is actuated through measurement, evaluation, and corrective action. Measurement is the process of measuring the relationship between planned performance and actual performance with respect to project objectives. The variables to be measured, the measurement scales, and the measuring approaches should be clearly specified during the planning stage. Corrective actions may involve rescheduling, the reallocation of resources, or the expedition of task performance. Control involves

- Tracking and reporting
- Measurement and evaluation
- Corrective action (plan revision, rescheduling, updating)

Project termination

Termination is the last stage of a project. The phase-out of a project is as important as its initiation. The termination of a project should be implemented expeditiously; it should not be allowed to drag on after the expected completion time. A terminal activity should be defined for the project during the planning phase. An example of a terminal activity may be the submission of a final report, the powering on of new equipment, or the signing of a release order. The conclusion of such an activity should be viewed as the completion of the project. Arrangements may be made for follow-up activities that may improve or extend the outcome of the project. These follow-up or spin-off projects should be managed as new projects but with proper input–output relationships within the sequence of projects.

Project systems implementation outline

The traditional project management framework encompasses the following broad sequence of categories (Badiru et al. 2008):

Planning → Organizing → Scheduling → Control → Termination

An outline of the functions to be carried out during a project should be made during the planning stage. A model of this outline follows. It may be necessary to rearrange the contents of the outline to fit the specific needs of the project.

Planning

1. Specify the project background.
   a. Define the current situation and process.
      i. Understand the process.
      ii. Identify important variables.
      iii. Quantify variables.

    b. Identify areas for improvement.
        i. List and discuss the areas.
        ii. Study potential strategies for a solution.
2. Define unique terminologies relevant to the project.
    a. Industry-specific terminologies.
    b. Company-specific terminologies.
    c. Project-specific terminologies.
3. Define the project goal and objectives.
    a. Write a mission statement.
    b. Solicit inputs and ideas from personnel.
4. Establish performance standards.
    a. Schedule.
    b. Performance.
    c. Cost.
5. Conduct a formal project feasibility study.
    a. Determine the impact on cost.
    b. Determine the impact on organization.
    c. Determine the project deliverables.
6. Secure management support.

Organizing

1. Identify the project management team.
    a. Specify the project organization structure.
        i. Matrix structure.
        ii. Formal and informal structures.
        iii. Justify the structure.
    b. Specify the departments involved and key personnel.
        i. Purchasing.
        ii. Materials management.
        iii. Engineering, design, manufacturing, and so on.
    c. Define the project management responsibilities.
        i. Select the project manager.
        ii. Write the project charter.
        iii. Establish the project policies and procedures.
2. Implement the Triple C model.
    a. Communication.
        i. Determine communication interfaces.
        ii. Develop a communication matrix.
    b. Cooperation.
        i. Outline the cooperation requirements, policies, and procedures.

  c. Coordination.
    i. Develop a work breakdown structure.
    ii. Assign task responsibilities.
    iii. Develop a responsibility chart.

Scheduling (resource allocation)

1. Develop a master schedule.
  a. Estimate the task duration.
  b. Identify task precedence requirements.
    i. Technical precedence.
    ii. Resource-imposed precedence.
    iii. Procedural precedence.
  c. Use analytical models.
    i. Critical path method (CPM).
    ii. PERT.
    iii. Gantt chart.
    iv. Optimization models.

Control (tracking, reporting, and correction)

1. Establish guidelines for tracking, reporting, and control.
  a. Define data requirements.
    i. Data categories.
    ii. Data characterization.
    iii. Measurement scales.
  b. Develop data documentation.
    i. Data update requirements.
    ii. Data quality control.
    iii. Establish data security measures.
2. Categorize the control points.
  a. Schedule an audit.
    i. Activity network and Gantt charts.
    ii. Milestones.
    iii. Delivery schedule.
  b. Performance audit.
    i. Employee performance.
    ii. Product quality.
  c. Cost audit.
    i. Cost containment measures.
    ii. Percent completion versus budget depletion.

3. Identify the implementation process.
   a. Comparison with targeted schedules.
   b. Corrective course of action.
      i. Rescheduling.
      ii. Reallocation of resources.

Termination (close, phase-out)

1. Conduct a performance review.
2. Develop a strategy for follow-up projects.
3. Arrange for personnel retention, release, and reassignment.

Documentation

1. Document the project outcome.
2. Submit the final report.
3. Archive the report for future reference.

Systems decision analysis

Systems decision analysis facilitates proper consideration of the essential elements of decisions in a project systems environment. These essential elements include the problem statement, the data and information requirements, the performance measure, the decision model, and the implementation of the decision. The recommended steps are described in the following sections.

Step 1. Problem statement

Solving a problem involves choosing between competing, and probably conflicting, alternatives. The components of problem solving in project management include

- Describing the problem (goals, performance measures)
- Defining a model to represent the problem
- Solving the model
- Testing the solution
- Implementing and maintaining the solution

Problem definition is very crucial. In many cases, the *symptoms* of a problem are more readily recognized than its *cause* and *location*. Even after the problem is accurately identified and defined, a benefit–cost analysis may be needed to determine if the cost of solving the problem is justified.

Step 2. Data and information requirements

Information is the driving force of the project decision process. Information clarifies the relative states of past, present, and future events. The collection, storage, retrieval, organization, and processing of raw data are important components of information generation. Without data, there can be no information. Without good information, there cannot be a valid decision. The essential requirements for generating information are

- Ensuring that an effective data collection procedure is followed
- Determining the type and the appropriate amount of data to collect
- Evaluating the data collected with respect to information potential
- Evaluating the cost of collecting the required data

For example, suppose a manager is presented with a recorded fact that says, "Sales for the last quarter are 10,000 units." This constitutes ordinary data. There are many ways of using these data to make a decision depending on the manager's value system. An analyst, however, can ensure the proper use of the data by transforming it into information, such as "Sales of 10,000 units for last quarter are within $x\%$ of the targeted value." This type of information is more useful to the manager for decision making.

Step 3. Performance measure

A performance measure for the competing alternatives should be specified. The decision maker assigns a perceived worth or value to the available alternatives. Setting a measure of performance is crucial to the process of defining and selecting alternatives. Some performance measures commonly used in project management are project cost, completion time, resource usage, and stability in the workforce.

Step 4. Decision model

A decision model provides the basis for the analysis and synthesis of information and is the mechanism by which competing alternatives are compared. To be effective, a decision model must be based on a systematic and logical framework for guiding project decisions. A decision model can be a verbal, graphical, or mathematical representation of the ideas in the decision-making process. A project decision model should have the following characteristics:

- A simplified representation of the actual situation
- An explanation and prediction of the actual situation
- Validity and appropriateness
- Applicability to similar problems

The formulation of a decision model involves three essential components.

- *Abstraction*: Determining the relevant factors
- *Construction*: Combining the factors into a logical model
- *Validation*: Assuring that the model adequately represents the problem

The basic types of decision models for project management are as follows:

- *Descriptive models*. These models are directed at describing a decision scenario and identifying the associated problem. For example, a project analyst might use a CPM network model to identify bottleneck tasks in a project.
- *Prescriptive models*. These models furnish procedural guidelines for implementing actions. The Triple C approach (Badiru, 2008), for example, is a model that prescribes the procedures for achieving *communication*, *cooperation*, and *coordination* in a project environment.
- *Predictive models*. These models are used to predict future events in a problem environment. They are typically based on historical data about the problem situation. For example, a regression model based on past data may be used to predict future productivity gains associated with expected levels of resource allocation. Simulation models can be used when uncertainties exist in the task durations or resource requirements.
- *Satisficing models*. These are models that provide trade-off strategies for achieving a satisfactory solution to a problem within given constraints. Goal programming and other multicriteria techniques provide good satisficing solutions. For example, these models are helpful in cases where time limitations, resource shortages, and performance requirements constrain the implementation of a project.
- *Optimization models*. These models are designed to find the best available solution to a problem, subject to a certain set of constraints. For example, a linear programming model can be used to determine the optimal product mix in a production environment.

In many situations, two or more of the preceding models may be involved in the solution of a problem. For example, a descriptive model might provide insights into the nature of the problem; an optimization model might provide the optimal set of actions to take in solving the problem; a satisficing model might temper the optimal solution with reality; a prescriptive model might suggest the procedures for implementing the selected solution; and a predictive model might present a projection of what to expect in the future.

Step 5. Making the decision

Using the available data, information, and decision model, the decision maker will determine the real-world actions that are needed to solve the stated problem. A sensitivity analysis may be useful for determining what changes in parameter values might cause a change in the decision.

Step 6. Implementing the decision

A decision represents the selection of an alternative that satisfies the objective stated in the problem statement. A good decision is useless until it is implemented. An important aspect of a decision is to specify how it is to be implemented. Selling the decision and the project to management requires a well-organized, persuasive presentation. The way a decision is presented can directly influence whether or not it is adopted. The presentation of a decision should include at least the following: an executive summary, the technical aspects of the decision, the managerial aspects of the decision, the resources required to implement the decision, the cost of the decision, the time frame for implementing the decision, and the risks associated with the decision.

Group decision making

Systems decisions are often complex, diffuse, distributed, and poorly understood. No one person has all the information to make all decisions accurately. As a result, crucial decisions are made by a group of people. Some organizations use outside consultants with the appropriate expertise to make recommendations for important decisions. Other organizations set up their own internal consulting groups without having to go outside the organization. Decisions can be made through linear responsibility; in which case, one person makes the final decision based on inputs from other people. Decisions can also be made through shared responsibility; in which case, a group of people share the responsibility for making joint decisions. The major advantages of group decision making are as follows:

1. The facilitation of a systems view of the problem environment.
2. The ability to share experience, knowledge, and resources. Many heads are better than one. A group will possess greater collective ability to solve a given decision problem.
3. Increased credibility. Decisions made by a group of people often carry more weight in an organization.
4. Improved morale. Personnel morale can be positively influenced because many people have the opportunity to participate in the decision-making process.

5. Better rationalization. The opportunity to observe other people's views can lead to an improvement in an individual's reasoning process.
6. The ability to accumulate more knowledge and facts from diverse sources.
7. Access to broader perspectives spanning different problem scenarios.
8. The ability to generate and consider alternatives from different perspectives.
9. The possibility of broad-based involvement, leading to a higher likelihood of support.
10. The possibility of group leverage for networking, communication, and political clout.

In spite of the much-desired advantages, group decision making does pose the risk of flaws. Some possible disadvantages of group decision making are as follows:

1. Difficulty in arriving at a decision
2. A slow operating time frame
3. The possibility of individuals having conflicting views and objectives
4. The reluctance of some individuals in implementing the decision
5. The potential for power struggles and conflicts within the group
6. A loss of productive employee time
7. Too much compromise, leading to less than optimal group output
8. The risk of one individual dominating the group
9. Overreliance on the group process, impeding the agility of management to make fast decisions
10. The risk of people dragging their feet due to repeated and iterative group meetings

Brainstorming

Brainstorming is a way of generating many new ideas. In brainstorming, the decision group comes together to discuss alternate ways of solving a problem. The members of the brainstorming group may be from different departments, may have different backgrounds and training, and may not even know one another. The diversity of the participants helps create a stimulating environment for generating different ideas from different viewpoints. The technique encourages the free outward expression of new ideas, no matter how far fetched they might appear. No criticism of any new idea is permitted during the brainstorming session.

A major concern in brainstorming is that extroverts may take control. For this reason, an experienced and respected individual should manage the discussions. The group leader establishes the procedure for proposing ideas, keeps the discussions in line with the group's mission, discourages disruptive statements, and encourages the participation of all members.

After the group has run out of ideas, open discussions are held to weed out the unsuitable ones. It is to be expected that even rejected ideas may eventually stimulate the generation of other more favorable ideas. Guidelines for improving brainstorming sessions are as follows:

- Focus on a specific decision problem.
- Keep ideas relevant to the intended decision.
- Be receptive to all new ideas.
- Evaluate the ideas on a relative basis after exhausting new ideas.
- Maintain an atmosphere conducive to cooperative discussions.
- Maintain a record of the ideas generated.

Delphi method

The traditional approach to group decision making is to obtain the opinion of experienced participants through open discussions. An attempt is made to reach a consensus among the participants. However, open group discussions are often biased because of the influence of subtle intimidation from dominant individuals. Even when the threat of a dominant individual is not present, opinions may still be swayed by group pressure. This is called the *bandwagon effect* of group decision making.

The Delphi method attempts to overcome these difficulties by requiring individuals to present their opinions anonymously through an intermediary. The method differs from other interactive group methods because it eliminates face-to-face confrontations. It was originally developed for forecasting applications, but it has been modified in various ways for application to different types of decision making. The method can be quite useful for project management decisions. It is particularly effective when decisions must be based on a broad set of factors. The Delphi method is normally implemented as follows:

1. *Problem definition*. A decision problem that is considered significant is identified and clearly described.
2. *Group selection*. An appropriate group of experts or experienced individuals is formed to address the particular decision problem. Both internal and external experts may be involved in the Delphi process. A leading individual is appointed to serve as the administrator of the decision process. The group may operate through correspondence or gather together in a room. In either case, all opinions are expressed anonymously. If the group meets in the same room, care should be taken to provide enough room so that each member does

not have the feeling that someone may accidentally or deliberately observe their responses.
3. *Initial opinion poll*. The technique is initiated by describing the problem to be addressed in unambiguous terms. The group members are requested to submit a list of major concerns in their specialty areas as they relate to the decision problem.
4. *Questionnaire design and distribution*. Questionnaires are prepared to address the areas of concern related to the decision problem. The written responses to the questionnaires are collected and organized by the administrator. The administrator aggregates the responses in a statistical format. For example, the average, mode, and median of the responses may be computed. This analysis is distributed to the decision group. Each member can then see how his or her responses compare with the anonymous views of the other members.
5. *Iterative balloting*. Additional questionnaires based on the previous responses are passed to the members. The members submit their responses again. They may choose to alter or not to alter their previous responses.
6. *Silent discussions and consensus*. The iterative balloting may involve anonymous written discussions of why some responses are correct or incorrect. The process is continued until a consensus is reached. A consensus may be declared after five or six iterations of the balloting or when a specified percentage (e.g., 80%) of the group agrees on the questionnaires. If a consensus cannot be declared on a particular point, it may be displayed to the whole group with a note that it does not represent a consensus.

In addition to its use in technological forecasting, the Delphi method has been widely used in other general decision making. Its major characteristics—the anonymity of responses, the statistical summary of responses, and the controlled procedure—make it a reliable mechanism for obtaining numeric data from subjective opinion. The major limitations of the Delphi method are as follows:

1. Its effectiveness may be limited in cultures where strict hierarchy, seniority, and age influence decision-making processes.
2. Some experts may not readily accept the contribution of nonexperts to the group decision-making process.
3. Since opinions are expressed anonymously, some members may take the liberty of making ludicrous statements. However, if the group composition is carefully reviewed, this problem may be avoided.

Nominal group technique

The nominal group technique is a silent version of brainstorming. It is a method of reaching consensus. Rather than asking people to state their ideas aloud, the team leader asks each member to jot down a minimum number of ideas—for example, five or six. A single list of ideas is then written on a chalkboard for the whole group to see. The group then discusses the ideas and weeds out some iteratively until a final decision is made. The nominal group technique is easier to control. Unlike brainstorming, where members may get into shouting matches, the nominal group technique permits members to silently present their views. In addition, it allows introverted members to contribute to the decision without the pressure of having to speak out too often.

In all of the group decision-making techniques, an important aspect that can enhance and expedite the decision-making process is to require that members review all pertinent data before coming to the group meeting. This will ensure that the decision process is not impeded by trivial preliminary discussions. Some disadvantages of group decision making are as follows:

1. Peer pressure in a group situation may influence a member's opinions or discussions.
2. In a large group, some members may not get to participate effectively in the discussions.
3. A member's relative reputation in the group may influence how well his or her opinion is rated.
4. A member with a dominant personality may overwhelm other members in the discussions.
5. The limited time available to the group may create pressure that forces some members to present their opinions without fully evaluating the ramifications of the available data.
6. It is often difficult to get all members of a decision group together at the same time.

Despite the noted disadvantages, group decision making has many definite advantages that may nullify the shortcomings. The advantages as presented earlier will have varying levels of effect from one organization to another. The Triple C approach (Badiru, 2008) may also be used to improve the success of decision teams. Teamwork can be enhanced in group decision making by adhering to the following guidelines:

1. Get a willing group of people together.
2. Set an achievable goal for the group.
3. Determine the limitations of the group.

4. Develop a set of guiding rules for the group.
5. Create an atmosphere conducive to group synergism.
6. Identify the questions to be addressed in advance.
7. Plan to address only one topic per meeting.

For major decisions and long-term group activities, arrange for team training that allows the group to learn the decision rules and responsibilities together. The steps for the nominal group technique are

1. Silently generate ideas in writing.
2. Record ideas without discussion.
3. Conduct group discussion for the clarification of meaning, not argument.
4. Vote to establish the priority or rank of each item.
5. Discuss the vote.
6. Cast the final vote.

Interviews, surveys, and questionnaires

Interviews, surveys, and questionnaires are important information-gathering techniques. They also foster cooperative working relationships. They encourage direct participation and inputs into project decision-making processes. They provide an opportunity for employees at the lower levels of an organization to contribute ideas and inputs for decision making. The greater the number of people involved in the interviews, surveys, and questionnaires, the more valid the final decision. The following guidelines are useful for conducting interviews, surveys, and questionnaires to collect data and information for project decisions.

1. Collect and organize background information and supporting documents on the items to be covered by the interview, survey, or questionnaire.
2. Outline the items to be covered and list the major questions to be asked.
3. Use a suitable medium of interaction and communication: telephone, fax, e-mail, face to face, observation, meeting venue, poster, or memo.
4. Tell the respondent the purpose of the interview, survey, or questionnaire, and indicate how long it will take.
5. Use open-ended questions that stimulate ideas from the respondents.
6. Minimize the use of yes/no questions.
7. Encourage expressive statements that indicate the respondent's views.

8. Use the *who, what, where, when, why, and how* approach to elicit specific information.
9. Thank the respondents for their participation.
10. Let the respondents know the outcome of the exercise.

Multivoting

Multivoting is a series of votes used to arrive at a group decision. It can be used to assign priorities to a list of items. It can be used at team meetings after a brainstorming session has generated a long list of items. Multivoting helps reduce such long lists to a few items, usually three to five. The steps for multivoting are

1. Take a first vote. Each person votes as many times as desired, but only once per item.
2. Circle the items receiving a relatively higher number of votes than the other items (i.e., a majority vote).
3. Take a second vote. Each person votes for a number of items equal to one-half the total number of items circled in step 2. Only one vote per item is permitted.
4. Repeat steps 2 and 3 until the list is reduced to three to five items depending on the needs of the group. It is not recommended to multivote down to only one item.
5. Perform further analysis of the items selected in step 4, if needed.

Project systems hierarchy

To reemphasize the general systems hierarchy presented in Chapter 1, this section discusses systems hierarchy within the specific context of project systems hierarchy.

The traditional concepts of systems analysis are applicable to the project process. The definitions of a project system and its components are as follows:

- *System*. A project system consists of interrelated elements organized for the purpose of achieving a common goal. The elements are organized to work synergistically to generate a unified output that is greater than the sum of the individual outputs of the components.
- *Program*. A program is a very large and prolonged undertaking. Such endeavors often span several years. Programs are usually associated with particular systems. For example, we may have a space exploration program within a national defense system.

- *Project*. A project is a time-phased effort of much smaller scope and duration than a program. Programs are sometimes viewed as consisting of a set of projects. Government projects are often called *programs* because of their broad and comprehensive nature. Industry tends to use the term *project* because of the short-term and focused nature of most industrial efforts.
- *Task*. A task is a functional element of a project. A project is composed of a sequence of tasks that all contribute to the overall project goal.
- *Activity*. An activity can be defined as a single element of a project. Activities are generally smaller in scope than tasks. In a detailed analysis of a project, an activity may be viewed as the smallest, most practically indivisible work element of the project. For example, we can regard a manufacturing plant as a system. A plant-wide endeavor to improve productivity can be viewed as a program. The installation of a flexible manufacturing system is a project within the productivity improvement program. The process of identifying and selecting equipment vendors is a task, and the actual process of placing an order with a preferred vendor is an activity.

The emergence of systems development has had an extensive effect on project management in recent years. A system can be defined as a collection of interrelated elements brought together to achieve a specified objective. In a management context, the purposes of a system are to develop and manage operational procedures and to facilitate an effective decision-making process. Some of the most common characteristics of a system include

1. Interaction with the environment
2. Objectives
3. Self-regulation
4. Self-adjustment

Representative components of a project system are the organizational, planning, scheduling, information management, control, and project delivery subsystems. The primary responsibilities of project analysts involve ensuring the proper flow of information throughout the project system. The classical approach to the decision process follows rigid lines of organizational charts. By contrast, the systems approach considers all the interactions necessary among the various elements of an organization in the decision process.

The various elements (or subsystems) of the organization act simultaneously in a separate but interrelated fashion to achieve a common goal. This synergism helps to expedite the decision process and to enhance the effectiveness of decisions. The supporting commitments from other subsystems of the organization serve to counterbalance the weaknesses of a given subsystem. Thus, the overall effectiveness of the system is greater than the sum of the individual results from the subsystems.

The increasing complexity of organizations and projects makes the systems approach essential management environment. As the number of complex projects increase, there will be an increasing need for project management professionals who can function as systems integrators. Project management techniques can be applied to the various stages of implementing a system as shown in the following guidelines:

1. *Systems definition*: Define the system and associated problems using keywords that signify the importance of the problem to the overall organization. Locate experts in this area who are willing to contribute to the effort. Prepare and announce the development plan.
2. *Personnel assignment*: The project group and the respective tasks should be announced, a qualified project manager should be appointed, and a solid line of command should be established and enforced.
3. *Project initiation*: Arrange an organizational meeting, during which a general approach to the problem should be discussed. Prepare a specific development plan and arrange for the installation of the required hardware and tools.
4. *System prototype*: Develop a prototype system, test it, and learn more about the problem from the test results.
5. *Full system development*: Expand the prototype to a full system, evaluate the user interface structure, and incorporate user training facilities and documentation.
6. *System verification*: Involve experts and potential users, ensure that the system performs as designed, and debug the system as needed.
7. *System validation*: Ensure that the system yields the expected outputs. Validate the system by evaluating performance levels, such as the percentage of success in so many trials, measuring the level of deviation from expected outputs, and measuring the effectiveness of the system output in solving the problem.
8. *System integration*: Implement the full system as planned, ensure the system can coexist with systems already in operation, and arrange for technology transfer to other projects.
9. *System maintenance*: Arrange for the continuing maintenance of the system. Update solution procedures as new pieces of information become available. Retain responsibility for system performance or delegate to well-trained and authorized personnel.
10. *Documentation*: Prepare full documentation of the system, prepare a user's guide, and appoint a user consultant.

Systems integration permits the sharing of resources, such as physical equipment, concepts, information, and skills. Systems integration is now a major concern of many organizations. Even some of the organizations that traditionally compete and typically shun cooperative efforts are beginning to appreciate the value of integrating their operations. For these reasons, systems integration has emerged as a major interest in business. Systems integration may involve the physical integration of technical components, the objective integration of operations, the conceptual integration of management processes, or a combination of any of these.

Systems integration involves the linking of components to form subsystems and the linking of subsystems to form composite systems within a single department and/or across departments. It facilitates the coordination of technical and managerial efforts to enhance organizational functions, reduce cost, save energy, improve productivity, and increase the utilization of resources. Systems integration emphasizes the identification and coordination of the interface requirements among the components in an integrated system. The components and subsystems operate synergistically to optimize the performance of the total system. Systems integration ensures that all performance goals are satisfied with a minimum expenditure of time and resources. Integration can be achieved in several forms including the following:

1. *Dual-use integration*: This involves the use of a single component by separate subsystems to reduce both the initial cost and the operating cost during the project life cycle.
2. *Dynamic resource integration*: This involves integrating the resource flows of two normally separate subsystems so that the resource flow from one to or through the other minimizes the total resource requirements in a project.
3. *Restructuring of functions*: This involves the restructuring of functions and the reintegration of subsystems to optimize costs when a new subsystem is introduced into the project environment.

Systems integration is particularly important when introducing new technology into an existing system. It involves coordinating new operations to coexist with existing operations. It may require the adjustment of functions to permit the sharing of resources, the development of new policies to accommodate product integration, or the realignment of managerial responsibilities. It can affect both the hardware and software components of an organization. The following guidelines and questions are relevant to systems integration.

- What are the unique characteristics of each component in the integrated system?
- How do the characteristics complement one another?
- What physical interfaces exist among the components?

- What data/information interfaces exist among the components?
- What ideological differences exist among the components?
- What are the data flow requirements for the components?
- Are there similar integrated systems operating elsewhere?
- What are the reporting requirements in the integrated system?
- Are there any hierarchical restrictions on the operations of the components of the integrated system?
- What internal and external factors are expected to influence the integrated system?
- How can the performance of the integrated system be measured?
- What benefit–cost documentations are required for the integrated system?
- What is the cost of designing and implementing the integrated system?
- What are the relative priorities assigned to each component of the integrated system?
- What are the strengths of the integrated system?
- What are the weaknesses of the integrated system?
- What resources are needed to keep the integrated system operating satisfactorily?
- Which section of the organization will have primary responsibility for the operation of the integrated system?
- What are the quality specifications and requirements for the integrated systems?

The integrated approach to project management starts with a managerial analysis of the project effort. Goals and objectives are defined, a mission statement is written, and the statement of work is developed. After these, traditional project management approaches, such as the selection of an organization structure, are employed. Conventional analytical tools, including the CPM and the precedence diagramming method (PDM), are then mobilized. The use of optimization models is then appropriate. Some of the parameters to be optimized are cost, resource allocation, and schedule length. It should be understood that not all project parameters will be amenable to optimization. Commercial project management software should only be used after the managerial functions have been completed. Some project management software has built-in capabilities for planning and optimization needs.

A frequent mistake in project management is the rush to use project management software without first completing the planning and analytical studies required by the project. Project management software should be used as a management tool, the same way a word processor is used as a writing tool. Using a word processor will be ineffective without first organizing one's thoughts about what is to be written. Project management is much more than just the software. If project management is carried out in accordance with the integration approach presented in the flowchart, the odds of success will be increased. Of course, the structure of the flowchart should not be rigid. Flows and interfaces among the blocks in the flowchart may need to be altered or modified depending on specific project needs.

Work breakdown structure

Work breakdown structure (WBS) refers to the itemization of a project for planning, scheduling, and control purposes. It presents the inherent components of a project in a structured block diagram or interrelationship flow chart. A WBS shows the relative hierarchies of the parts (phases, segments, milestone, etc.) of a project. The purpose of constructing a WBS is to analyze the elemental components of the project in detail. If a project is properly designed through the application of a WBS at the project planning stage, it becomes easier to estimate the cost and time requirements of a project. Project control is also enhanced by the ability to identify how the components of a project link together. Tasks that are contained in the WBS collectively describe the overall project goal. Overall project planning and control can be improved by using a WBS approach. A large project may be broken down into smaller subprojects that may, in turn, be systematically broken down into task groups. Thus, a WBS permits the implementation of a *divide and conquer* concept for project control.

Individual components in a WBS are referred to as WBS elements, and the hierarchy of each is designated by a level identifier. Elements at the same level of subdivision are said to be of the same WBS level. Descending levels provide increasingly detailed definitions of project tasks. The complexity of a project and the degree of control desired determine the number of levels in the WBS. Each component is successively broken down into smaller details at lower levels. The process may continue until specific project activities are reached. In effect, the structure of the WBS looks very much like an organizational chart. The basic approach for preparing a WBS is as follows:

- *Level 1 WBS*: This contains only the final goal of the project. This item should be identifiable directly as an organizational budget item.
- *Level 2 WBS*: This level contains the major subsections of the project. These subsections are usually identified by their contiguous location or by their related purposes.
- *Level 3 WBS*: Level 3 of the WBS structure contains the definable components of the level 2 subsections. In technical terms, this may be referred to as the *finite element* level of the project.

The subsequent levels of a WBS are constructed in more specific detail, depending on the span of control desired. If a complete WBS becomes too crowded, separate WBS layouts may be drawn for the level 2 components. A statement of work (SOW) or WBS summary should accompany the WBS. The SOW is a narrative of the work to be done. It should include the objectives of the work, its scope, resource requirements, tentative due date, feasibility statements, and so on. A good analysis of the WBS structure will make it easier to perform resource work rate analysis.

Work feasibility

The feasibility of a project can be ascertained in terms of technical factors, economic factors, or both. A feasibility study is documented with a report showing all the ramifications of the project and should be broken down into the following categories:

Technical feasibility: Technical feasibility refers to the ability of the process to take advantage of the current state of the technology in pursuing further improvement. The technical capability of the personnel as well as the capability of the available technology should be considered.

Managerial feasibility: Managerial feasibility involves the capability of the infrastructure of a process to achieve and sustain process improvement. Management support, employee involvement, and commitment are key elements required to ascertain managerial feasibility.

Economic feasibility: This involves the ability of the proposed project to generate economic benefits. A benefit–cost analysis and a break-even analysis are important aspects of evaluating the economic feasibility of new industrial projects. The tangible and intangible aspects of a project should be translated into economic terms to facilitate a consistent basis for evaluation.

Financial feasibility: Financial feasibility should be distinguished from economic feasibility. Financial feasibility involves the capability of the project organization to raise the appropriate funds needed to implement the proposed project. Project financing can be a major obstacle in large multiparty projects because of the level of capital required. Loan availability, credit worthiness, equity, and loan schedule are important aspects of financial feasibility analyses.

Cultural feasibility: Cultural feasibility deals with the compatibility of the proposed project with the cultural setup of the project environment. In labor-intensive projects, planned functions must be integrated with the local cultural practices and beliefs. For example, religious beliefs may influence what an individual is willing or not willing to do.

Social feasibility: Social feasibility addresses the influences that a proposed project may have on the social system in the project environment. The ambient social structure may be such that certain categories of workers may be in short supply or nonexistent. The effect of the project on the social status of the project participants must be assessed to ensure compatibility. It should be recognized that workers in certain industries may have certain status symbols within the society.

Safety feasibility: Safety feasibility is another important aspect that should be considered in project planning. Safety feasibility refers to an analysis of whether the project is capable of being implemented and operated safely, with minimal adverse effects on the environment. Unfortunately, environmental impact assessments are often not adequately addressed in complex projects. As an example, the North America Free Trade Agreement (NAFTA) between the United States, Canada, and Mexico was temporarily suspended in 1993 because of legal considerations regarding the potential environmental impacts of the projects to be undertaken under the agreement.

Political feasibility: A politically feasible project may be referred to as a *politically correct project*. Political considerations often dictate the direction for a proposed project. This is particularly true for large projects with national visibility that may have significant government inputs and political implications. For example, political necessity may be a source of support for a project regardless of the project's merits. On the other hand, worthy projects may face insurmountable opposition simply because of political factors. Political feasibility analysis requires an evaluation of the compatibility of the project goals with the prevailing goals of the political system. In general, a feasibility analysis for a project should include the following items:

1. *Need analysis*: This indicates the recognition of a need for the project. The need may affect the organization itself, another organization, the public, or the government. A preliminary study is conducted to confirm and evaluate the need. A proposal of how the need may be satisfied is then made. Pertinent questions that should be asked include the following:
   - Is the need significant enough to justify the proposed project?
   - Will the need still exist by the time the project is completed?
   - What are the alternative means of satisfying the need?
   - What are the economic, social, environmental, and political impacts of the need?
2. *Process work*: This is the preliminary analysis done to determine what will be required to satisfy the need. The work may be performed by a consultant who is an expert in the project field. The

preliminary study often involves system models or prototypes. For technology-oriented projects, artist conceptions and scale models may be used for illustrating the general characteristics of a process. A simulation of the proposed system can be carried out to predict the outcome before the actual project starts.

3. *Engineering and design*: This involves a detailed technical study of the proposed project. Written quotations are obtained from suppliers and subcontractors as needed. Technology capabilities are evaluated as needed. Product design, if needed, should be done at this stage.
4. *Cost estimate*: This involves estimating project cost to an acceptable level of accuracy. Levels of around –5% to +15% are common at this level of a project plan. Both the initial and operating costs are included in the cost estimation. Estimates of capital investment and recurring and nonrecurring costs should also be contained in the cost estimate document. A sensitivity analysis can be carried out on the estimated cost values to see how sensitive the project plan is to changes in the project scenario.
5. *Financial analysis*: This involves an analysis of the cash flow profile of the project. The analysis should consider rates of return, inflation, sources of capital, payback periods, the break-even point, residual values, and sensitivity.
6. *Project impacts*: This portion of the feasibility study provides an assessment of the impact of the proposed project. Environmental, social, cultural, political, and economic impacts may be some of the factors that will determine how a project is perceived by the public. The value-added potential of the project should also be assessed.
7. *Conclusions and recommendations*: The feasibility study should end with the overall outcome of the project analysis. This may constitute either an endorsement or a disapproval of the project.

Motivating the worker

Motivation is an essential component of implementing a project plan. Those who will play a direct role in the project must be motivated to ensure productive participation. Direct beneficiaries of the project must be motivated to make good use of the outputs of the project. Other groups must be motivated to play supporting roles to the project. Motivation may take several forms. For projects that are of a short-term nature, motivation could either be impaired or enhanced by the strategy employed. Impairment may occur if a participant views the project as a mere disruption of regular activities or as a job without long-term benefits. Long-term projects have the advantage of giving participants enough time to readjust to the project efforts.

Theory X principle of motivation

Theory X assumes that the worker is essentially uninterested and unmotivated to perform his or her work. Motivation must be instilled in the worker by the adoption of external motivating agents. A Theory X worker is inherently indolent and requires constant supervision and prodding to get him or her to perform. To motivate a Theory X worker, a mixture of managerial actions may be needed. Examples of motivation approaches under Theory X include

- Rewards to recognize improved effort
- Strict rules to constrain worker behavior
- Incentives to encourage better performance
- Threats to job security associated with performance failure

Theory Y principle of motivation

Theory Y assumes that the worker is naturally interested and motivated to perform his or her job. The worker views the job function positively and uses self-control and self-direction to pursue the project goals. Under Theory Y, management has the task of taking advantage of the worker's readiness and positive intuition so that his or her actions coincide with the project objectives. Thus, a Theory Y manager attempts to use the worker's self-direction as the principal instrument for accomplishing work. In general, Theory Y management encourages the following:

- Worker-designed job methodology
- Worker participation in decision making
- Cordial management–worker relationship
- Worker individualism within acceptable company limits

Motivating and demotivating factors

The Herzberg motivation concept takes a look at the characteristics of work itself as the motivating factor. There are two motivational factors, classified as *hygiene factors* and *motivators*. Hygiene factors are necessary but not sufficient conditions for a contented worker. The negative aspects of the factors may lead to a disgruntled worker, whereas the positive aspects do not necessarily enhance motivation. Examples include the following:

1. *Administrative policies*: Bad policies can lead to the discontent of workers, while good policies are viewed as routine, with no specific contribution to improving worker satisfaction.

2. *Supervision*: A bad supervisor can make a worker unhappy and less productive, but a good supervisor cannot necessarily improve worker performance.
3. *Working conditions*: Bad working conditions can enrage workers, but good working conditions do not automatically generate improved productivity.
4. *Salary*: Low salaries can make a worker unhappy, disruptive, and uncooperative, but a raise will not necessarily encourage him or her to perform better. While a raise in salary will not necessarily increase professionalism, a reduction in salary will most certainly have an adverse effect on morale.
5. *Personal life*: Miserable personal life can adversely affect a worker's performance, but a happy life does not imply that he or she will be a better worker.
6. *Interpersonal relationships*: Good peer, superior, and subordinate relationships are important to keep a worker happy and productive, but extraordinarily good relations do not guarantee that he or she will be more productive.
7. *Social and professional status*: Low status can force a worker to perform at his or her low "level," whereas high status does not imply that he or she will perform at a higher level.
8. *Security*: A safe environment may not motivate a worker to perform better, but unsafe conditions will certainly impede his or her productivity.

Motivators are motivating agents that should be inherent in the work itself. If necessary, project task assignments should be redesigned (or reengineered) to include inherent motivating factors. Some guidelines follow.

1. *Achievement*: The job design should facilitate opportunities for worker achievement and advancement toward personal goals.
2. *Recognition*: The mechanism for recognizing superior performance should be incorporated into the task assignment. Opportunities for recognizing innovation should be built into the task.
3. *Work content*: The work content should be interesting enough to motivate and stimulate the creativity of the worker. The amount of work and the organization of the work should be designed to fit a worker's needs.
4. *Responsibility*: The worker should have some measure of responsibility for how his or her job is performed. Personal responsibility leads to accountability, which leads to better performance.
5. *Professional growth*: The work should offer an opportunity for advancement so that the worker can set his own achievement level for professional growth within a project plan.

Management by objective

Management by objective (MBO) is a management concept whereby a worker is allowed to take responsibility for the design and performance of a task under controlled conditions. It gives each worker a chance to set his or her own objectives in achieving project goals. The worker can monitor his own progress and take corrective actions when needed without management intervention. Workers under the concept of Theory Y appear to be the best suited for the MBO concept. MBO has some disadvantages, however, which include the possible abuse of the freedom to self-direct and the possible disruption of overall project coordination. The advantages of MBO include the following:

1. Encouraging each worker to find better ways of performing the job
2. Avoiding the over-supervision of professionals
3. Helping workers become better aware of what is expected of them
4. Permitting timely feedback on worker performance

Management by exception

Management by exception (MBE) is an after-the-fact management approach to the issue of control. Contingency plans are not made and there is no rigid monitoring. Deviations from expectations are viewed as exceptions to the normal course of events. When intolerable deviations from plans occur, they are investigated and an action is taken. The major advantage of MBE is that it lessens the management workload and reduces the cost of management. However, it is a risky concept to follow, especially for high-stakes industry projects. Many of the problems that can develop in complex projects are such that after-the-fact corrections are expensive or impossible.

Matrix organization structure

The matrix organization is a frequently used organizational structure in industry. It is used where there are multiple managerial accountabilities and responsibilities for a project. It combines the advantages of the traditional structure and the product organization structure. The hybrid configuration of the matrix structure facilitates maximum resource utilization and increased performance within time, cost, and performance constraints. There are usually two chains of command involving both horizontal and vertical reporting lines. The horizontal line deals with the functional line of responsibility, while the vertical line deals with the project line of responsibility.

The advantages of matrix organization include the following:

- Good team interaction.
- The consolidation of objectives.
- A multilateral flow of information.
- Lateral mobility for job advancement.
- Individuals have an opportunity to work on a variety of projects.
- The efficient sharing and utilization of resources.
- Reduced project cost due to the sharing of personnel.
- The continuity of functions after project completion.
- Stimulating interactions with other functional teams.
- Functional lines rally to support the project efforts.
- Each person has a "home" office after project completion.
- The company knowledge base is equally available to all projects.

Some of the disadvantages of matrix organization are as follows:

- The matrix response time may be slow for fast-paced projects.
- Each project organization operates independently.
- Overhead costs due to additional lines of command.
- Potential conflicts of project priorities.
- Problems inherent in having multiple bosses.
- The complexity of the structure.

Traditionally, industrial projects are conducted in serial functional implementations such as research and development, engineering, manufacturing, and marketing. At each stage, unique specifications and work patterns may be used without consulting the preceding and succeeding phases. The consequence is that the end product may not possess the original intended characteristics. For example, the first project in the series might involve the production of one component, while the subsequent projects might involve the production of other components. The composite product may not achieve the desired performance because the components were not designed and produced from a unified point of view. The major appeal of matrix organization is that it attempts to provide synergy within the groups in an organization.

Team building of workers

Team building is a key part of project management, because workers are expected to perform not just on one team but on a multitude of interfacing teams. To facilitate effective team building, it is important to distinguish between a functional organization structure and an operational organization structure. Within a team, the functional structure is developed according to the functional lines of responsibility, while the operational structure takes into account the way workers actually organize themselves to accomplish a task. The operational organization can be centralized, hierarchical, or mixed.

In a centralized structure, workers rely on a specific individual or office for reporting and directive purposes. In a hierarchical structure, the lines of functional reporting existing in the organization are observed as workers pursue their tasks. This approach may discourage free and informal interaction, thereby creating obstacles to effective work. In a mixed structure, different heterogeneous teams of workers interact to solve a problem with the aid of a central coordination agent. The effectiveness of a team can be affected by its size and its purpose. Small teams are useful for responsiveness and prompt action. Large teams are useful for achieving a widespread information base, inclusion, and participation. Strategies for enhancing project team building include

1. Interoffice employee exchange programs
2. The interproject transferability of personnel
3. In-house training for new employees
4. The diversification and development of in-house job skills
5. The use of in-house personnel as in-house consultants

One of the most important ingredients for establishing a good team is commitment. The Triple C approach can help achieve such commitment. The organizational structure must create an environment that promotes, or even demands, teamwork. The project manager should strive to provide and maintain an atmosphere that fulfills the needs and expectations of workers. A team should offer opportunities for positive professional interactions as well as avenues for advancement. Some of the expectations and opportunities that a team environment should offer are as follows:

- Good team leadership
- A challenging (but not problematic) work environment
- Recognition for team accomplishment
- Recognition for individual contributions
- The ability to speak in a collective voice to management
- Opportunities for career growth
- Opportunities for social interaction
- The possibility of both exercising individual discretion and arriving at a collective consensus

Some of the barriers to team building are *changes in the work environment*, *a lack of member commitment*, *poor intrateam communication*, *a lack of management support*, and *the team superstar model*. On some teams, only one or a few individuals lead the efforts and get all the credit at the expense of the other team members. A teaming approach that allows everyone to pitch in and share credit equitably is preferred. If the team excels, everyone excels. Shared project glory is everyone's glory.

Project partnering

Project partnering involves having project teams and stakeholders operating as partners in the pursuit of project goals. The benefits of project partnering include improved efficiency, cost reduction, resource sharing, better effectiveness, increased potential for innovation, and the improved quality of products and services. Partnering creates a collective feeling of being together on the project. It fosters a positive attitude, which makes it possible to appreciate and accept the views of others. It also recognizes the objectives of all parties and promotes synergism. The communication, cooperation, and coordination concepts of the Triple C model facilitate partnering. Suggestions for setting up project partnering are as follows:

- Use an inclusive organization structure.
- Identify project stakeholders and clients.
- Create informational linkages.
- Identify the lead partner.
- Collate the objectives of partners.
- Use a responsibility chart to assign specific functions.

Team leadership

Good leaders lead by example. Others attempt to lead by dictating. It is an element of human nature that people learn and act best when good examples are available for them to emulate, and they generally do not forget the lessons they learn in this way. Good examples observed in childhood, for instance, can provide a lifetime's worth of guidance. A leader should have a spirit of performance that stimulates his or her subordinates to perform at their own best. Rather than dictating what needs to be done, a good leader should show what needs to be done. Showing, in this case, does not necessarily imply actual physical demonstration of what is to be done. Rather, it implies projecting a commitment to the function at hand and a readiness to participate as appropriate. In the traditional model, managers manage workers to get them to work. There may be no point of convergence or active participation. Modern managers, however, team up with workers to get the job done. A good leadership model will encompass listening and asking questions, specifying objectives, developing clear directions, removing obstacles, encouraging individual initiatives, learning from past experiences, reiterating the project requirements, and getting everyone involved.

Telling, showing, and involving are good practices for project team leadership. Project team leadership involves dealing with managers and supporting personnel across the functional lines of the project. It is a misconception to think that a leader leads only his or her own subordinates. Leadership responsibilities can cover functions vertically up or down. A good project leader can lead and inspire not only his or her subordinates but also the entire project organization, including the highest superiors. Generally, leadership involves recognizing an opportunity to make improvements in a project and taking the initiative to lead the implementation of the improvements. In addition to inherent personal qualities, leadership style can be influenced by training, experience, and dedication. Guidelines for project team leadership are as follows:

- Avoid organizational politics and personal egotism.
- Lead by example.
- Place principles above personality.
- Focus on the big picture of project goals.
- Build up credibility with successful leadership actions.
- Demonstrate integrity and ethics in decision processes.
- Cultivate and encourage a spirit of multilateral cooperation.
- Preach less and implement more.
- Back up words with action.

Activity scheduling

Project scheduling is the time-phased arrangement of project activities subject to precedence, time, and resource constraints in order to accomplish project objectives. Project scheduling is distinguished from industrial job shop, flow shop, and other production sequencing problems because of the nonrepetitive nature of most projects. In production scheduling, the scheduling problem follows a standard procedure that determines the characteristics of the production operations. A scheduling technique that works for one production run may be expected to work equally effectively for succeeding and identical production runs. In other words, reliable precedents can be found for production scheduling problems. On the other hand, projects are usually one-time endeavors that are rarely duplicated in identical circumstances. In some cases, it may be possible to duplicate the concepts of the whole project or a portion of it. The construction of a dam is a good example. The concepts of dam construction will, most likely, be the same in all dam operations. It would, however, be highly unlikely to have two dam construction projects that were exactly alike. Even if two projects are identical in many details, manpower availability (a critical component of any project) is likely to be different. This, of course, makes project scheduling more challenging than production scheduling.

Project scheduling represents the core of project management efforts because it involves the assignment of time periods to specific tasks within the work schedule. Resource availability, time limitations, urgency and priority, performance specification, precedence constraints, milestones, technical precedence constraints, and other factors complicate the scheduling process. Generally, project scheduling involves

- Analyzing resource availability in terms of human resources, materials, capital, and so on
- Applying scheduling techniques (CPM, PERT, Gantt charts, PDM)
- Tracking and reporting
- Control and termination

Work control

Project or work control requires that appropriate actions be taken to correct deviations from expected performance. Control involves measurement, evaluation, and correction. Measurement is the process of measuring the relationship between planned performance and actual performance with respect to project objectives. The variables to be measured, the measurement scales, and the measuring approaches should be clearly specified during the planning stage. Corrective actions may involve rescheduling, the reallocation of resources, or the expediting of tasks. In some cases, project termination is an element of project control.

The Triple C model

The Triple C model (Badiru, 2008) is an effective tool for project planning and control. The model can facilitate better resource management by identifying the crucial aspects of a project. The model states that project management can be enhanced by its implementation within the integrated functions of

- Communication
- Cooperation
- Coordination

The model facilitates a systematic approach to project planning, organizing, scheduling, and control. It highlights what must be done and when. It also helps to identify the resources (manpower, equipment, facilities, etc.) required for each effort. Triple C points out important questions, such as

- Does each project participant know what the objective is?
- Does each participant know his or her role in achieving the objective?
- What obstacles may prevent a participant from playing his or her role effectively?

Figure 10.3 shows a graphical representation of the Triple C model for project management in coordinated application with the DEJI model for work design. If resource management is viewed as a three-legged stool, then communication, cooperation, and coordination constitute the three legs. Communication channels provide the basis for effective communication, partnership forms the basis for cooperation, and organizational structure provides the basis for coordination. Consequently, there must be appropriate communication channels, partnership, and proper organizational structure for Triple C to be effective. This is summarized as follows:

1. For effective communication, create good communication channels.
2. For enduring cooperation, establish partnership arrangements.
3. For steady coordination, use a workable organization structure.

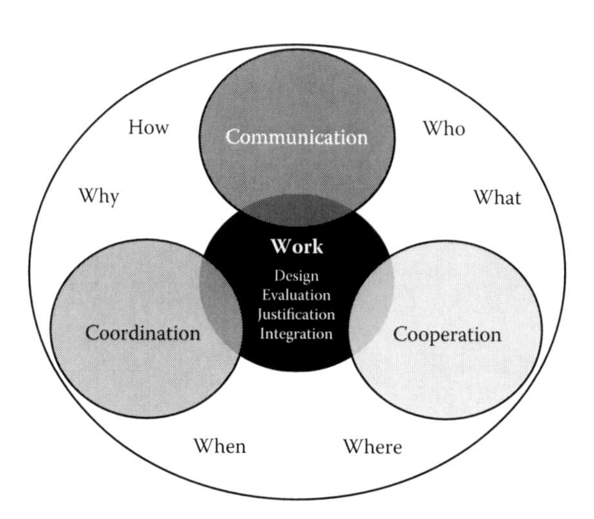

Figure 10.3 The Triple C approach combined with the DEJI model for work management.

Project communication

The communication function of project management involves making sure that all those concerned become aware of project requirements and progress. Those that will be affected by the project directly or indirectly, as direct participants or as beneficiaries, should be informed as appropriate regarding the following:

- The scope of the project
- The personnel contribution required
- The expected cost and merits of the project
- The project organization and implementation plan
- The potential adverse effects if the project should fail
- The alternatives, if any, for achieving the project goal
- The potential benefits (direct and indirect) of the project

Communication channels must be kept open throughout the project life cycle. In addition to internal communication, appropriate external sources should also be consulted. The project manager has several "must-do" requirements.

- Exude commitment to the project.
- Utilize a communication responsibility matrix.
- Facilitate multichannel communication interfaces.
- Identify internal and external communication needs.
- Resolve organizational and communication hierarchies.
- Encourage both formal and informal communication links.

When clear communication is maintained between management and employees and among peers, many project problems can be averted. Project communication may be carried out in one or more of the following formats:

- One-to-many
- One-to-one
- Many-to-one
- Written and formal
- Written and informal
- Oral and formal
- Oral and informal
- Nonverbal gestures

A communication responsibility matrix (table or spreadsheet) shows the links between sources of communication and targets of communication.

Cells within the matrix indicate the subject of the desired communication. There should be at least one filled cell in each row and each column of the matrix. This ensures that each individual within a department has at least one communication source or target associated with him or her. With a communication responsibility matrix, a clear understanding of what needs to be communicated and to whom can be developed.

Project cooperation

The cooperation of the project personnel must be explicitly elicited. Merely voicing consent for a project is not enough assurance of full cooperation. The participants and beneficiaries of the project must be convinced of its merits. Some of the factors that influence cooperation in a project environment include manpower requirements, resource requirements, budget limitations, past experiences, conflicting priorities, and a lack of uniform organizational support. A structured approach to seeking cooperation should clarify the following:

- The cooperative efforts required
- The precedents for future projects
- The implications of what a lack of cooperation can do to a project
- The criticality of cooperation to project success
- The organizational impact of cooperation
- The time frame involved in the project
- The rewards for good cooperation

Cooperation is a basic virtue of human interaction. More projects fail due to a lack of cooperation and commitment than any other project factors. To secure and retain the cooperation of project participants, their first reaction to the project must be positive. The most positive aspects of a project should be the first items of project communication. Guidelines for securing cooperation for projects are as follows:

- Establish achievable goals for the project.
- Clearly outline the individual commitments required.
- Integrate project priorities with existing priorities.
- Eliminate the fear of job loss due to automation.
- Anticipate and eliminate potential sources of conflict.
- Use an open-door policy to address project grievances.
- Remove skepticism by documenting the merits of the project.

For resource management, there are different types of cooperation that should be considered and encouraged. Some examples of these are as follows:

- *Functional cooperation*: This is cooperation induced by the nature of the functional relationship between two groups. The two groups may be required to perform related functions that can only be accomplished through mutual cooperation.
- *Social cooperation*: This is the type of cooperation effected by the social relationship between two groups. The prevailing social relationship motivates cooperation that may be useful in getting project work done.
- *Legal cooperation*: Legal cooperation is the type of cooperation that is imposed through some authoritative requirement. In this case, the participants may have no choice other than to cooperate.
- *Administrative cooperation*: This is cooperation brought on by administrative requirements that make it imperative that two groups work together on a common goal.
- *Associative cooperation*: This is a type of cooperation that may also be referred to as *collegiality*. The level of cooperation is determined by the association that exists between two groups.
- *Proximity cooperation*: Cooperation due to the fact that two groups are geographically close is referred to as proximity cooperation. Being close makes it imperative that the two groups work together.
- *Dependency cooperation*: This is cooperation caused by the fact that one group depends on another group for some important aspect. Such dependency is usually of a mutual, two-way nature. One group depends on the other for one thing, while the latter group depends on the former for some other thing.
- *Imposed cooperation*: In this type of cooperation, external agents must be employed to induce cooperation between two groups. This is applicable for cases where the two groups have no natural reason to cooperate. This is where the approaches presented earlier for seeking cooperation can become very useful.
- *Lateral cooperation*: Lateral cooperation involves cooperation with peers and immediate associates. Lateral cooperation is often easy to achieve because existing lateral relationships create a conducive environment for project cooperation.
- *Vertical cooperation*: Vertical or hierarchical cooperation refers to cooperation that is implied by the hierarchical structure of the project. For example, subordinates are expected to cooperate with their superiors.

Whichever type of cooperation is available in a project environment, the cooperative forces should be channeled toward achieving project goals. Documentation of the prevailing level of cooperation is useful for winning further support for a project. Clarification of project priorities will facilitate personnel cooperation. The relative priorities of multiple projects should be specified so that a project that is of high priority to one segment of an organization is also of high priority to all groups within the organization. Interestingly, competition can be used as a mechanism for cooperation. This happens in an environment where constructive competition paves the way for cooperation of the form that we can refer to as *coopetition*.

Project commitment

Cooperation must be supported with commitment. To cooperate is to support the ideas of a project. To commit is to willingly and actively participate in project efforts again and again throughout both the easy times and the vicissitudes of the project. The ungrudging provision of resources is one way that management can express commitment to a project.

Triple C + commitment = project success

By using a Pareto-type distribution, the cooperating elements of a project can be classified into the following three levels:

1. Top 10% (Easily cooperative)
2. Middle 80% (Good prospects for cooperation)
3. Bottom 10% (Lost causes)

In terms of where to place efforts for cooperation, the top 10% do not need much effort, while the bottom 10% do not deserve much effort. The top 10% are the motivated individuals who will easily cooperate (i.e., Theory Y workers), while the bottom 10% are the disagreeable individuals who will fail to see reason no matter what is presented to them (i.e., Theory X workers). The best way to deal with those bottom 10% is through accommodation or exclusion. Project efforts should be concentrated where the most gains can be achieved.

Project coordination

After successfully initiating the communication and cooperation functions, the efforts of the project personnel must be coordinated. Coordination facilitates the harmonious organization of the project efforts. The development of a responsibility chart can be very helpful at this stage. A responsibility chart is a matrix consisting of columns of individual or functional departments and rows of required actions. Cells within the matrix are filled with relationship codes that indicate who is responsible for what. The responsibility matrix helps to avoid neglecting crucial communication requirements and obligations. It can help resolve questions such as the following:

- Who is to do what?
- How long will it take?
- Who is to inform whom of what?
- Whose approval is needed for what?
- Who is responsible for which results?
- What personnel interfaces are required?
- What support is needed from whom and when?

Resolving project conflicts

When implemented as an integrated process, the Triple C model can help avoid conflicts in a project. When conflicts do develop, it can help in resolving the conflicts. Several types of conflicts can develop in the project environment. Some of these conflicts are as follows:

Scheduling conflicts can develop because of improper timing or sequencing of project tasks. This is particularly common in large multiple projects. Procrastination can lead to having too much to do at once, thereby creating a clash of project functions and discord between project team members. Inaccurate estimates of time requirements may lead to unfeasible activity schedules. Project coordination can help avoid schedule conflicts.

Cost conflicts. Project cost may not be generally acceptable to the clients of a project. This will lead to project conflict. Even if the initial cost of the project is acceptable, a lack of cost control during project implementation can lead to conflicts. Poor budget allocation approaches and the lack of a financial feasibility study will cause cost conflicts later on in a project. Communication and coordination can help prevent most of the adverse effects of cost conflicts.

Performance conflicts. If clear performance requirements are not established, performance conflicts will develop. A lack of clearly defined performance standards can lead each person to evaluate his or her own performance based on personal value judgments. In order to uniformly evaluate the quality of work and monitor project progress, performance standards should be established by using the Triple C approach.

Management conflicts. There must be a two-way alliance between management and the project team. The views of management should be understood by the team. The views of the team should be appreciated by management. If this does not happen, management conflicts will develop. A lack of a two-way interaction can lead to strikes and industrial action, which can be detrimental to project objectives. The Triple C approach can help create a conducive dialogue environment between management and the project team.

Technical conflicts. If the technical basis of a project is not sound, technical conflicts will develop. Manufacturing and automation projects are particularly prone to technical conflicts because of their significant dependence on technology. Lacking a comprehensive technical feasibility study will lead to technical conflicts. Performance requirements and systems specifications can be integrated through the Triple C approach to avoid technical conflicts.

Priority conflicts can develop if project objectives are not defined properly and applied uniformly across a project. A lack of direction in project definition can lead each project member to define individual goals that may be in conflict with the intended goal of a project. A lack of consistency in the project mission is another potential source of priority conflicts. The overassignment of responsibilities with no guidelines for relative significance levels can also lead to priority conflicts. Communication can help defuse priority conflicts.

Resource conflicts. Resource allocation problems are a major source of conflicts in project management. Competition for resources, including personnel, tools, hardware, software, and so on, can lead to disruptive clashes among project members. The Triple C approach can help secure resource cooperation.

Power conflicts. Project politics can lead to power plays as one individual seeks to widen his or her scope of power. This can, obviously, adversely affect the progress of a project. Project authority and project power should be clearly differentiated: project authority is the control that a person has by virtue of his or her functional post, while project power relates to the clout and influence a person can exercise due to connections within the administrative structure. People with popular personalities can often wield a lot of project power in spite of low or nonexistent project authority. The Triple C model can facilitate a positive marriage of project authority and power to the benefit of project goals. This will help define clear leadership for a project.

Personality conflicts are a common problem in projects involving a large group of people. The larger the project, the larger the size of the management team needed to keep things running. Unfortunately, a larger management team also creates opportunities for personality conflicts. Communication and cooperation can help defuse personality conflicts. Some guidelines for resolving project conflicts are as follows:

- Approach the source of conflict.
- Gather all the relevant facts.

- Notify those involved in writing.
- Solicit mediation.
- Report to the appropriate authorities.
- Use a grievance resolution program within the organization.

References

Badiru, A. B. (2008), *Triple C Model of Project Management: Communication, Cooperation, and Coordination*, CRC Press, Boca Raton, FL.

Badiru, A. B., S. A. Badiru, and I. A. Badiru (2008), *Industrial Project Management: Concepts, Tools, and Techniques*, CRC Press, Boca Raton, FL.

Badiru, I. A. (2016), Comments about work management, interview with an auto industry senior engineer about corporate views of work design, Beavercreek, OH, October 29.

Hammersmith, A. G. (2006), Implementing a PMO: The diplomatic pit bull, workshop presentation at East Tennessee PMI chapter meeting, January 10.

chapter eleven

Considerations for worker well-being*

Overview

Action to address workforce functioning and productivity requires a broader approach than the traditional scope of occupational safety and health. Focus on "well-being" may be one way to develop a more encompassing objective. Well-being is widely cited in public policy pronouncements, but often as "... and well-being" (e.g., health and well-being). It is generally not defined in policy and rarely operationalized for functional use. Many definitions of well-being exist in the occupational realm. Generally, it is a synonym for health and a summative term to describe a flourishing worker who benefits from a safe, supportive workplace, engages in satisfying work, and enjoys a fulfilling work life. We identified issues for considering well-being in public policy related to workers and the workplace.

Background

Major changes in population demographics and the world of work have significant implications for the workforce, business, and the nation (Stone, 2004; Howard, 2010; Kompier, 2006; Schulte, 2006; Holzer, 2007; Cummings, 2008; Kitt, 2013; Sparks et al., 2001). New patterns of hazards, resulting from the interaction of work and nonwork factors, are affecting the workforce. (Stone, 2004; Howard, 2010; Sparks et al., 2001; Schulte et al., 2012; Guillemin, 2011; Pandalai et al., 2013) As a consequence, there is a need for an overarching or unifying concept that can be operationalized to optimize the benefits of work and simultaneously address these overlapping hazards. Traditionally, the distinct disciplines of occupational safety and health, human resources, health promotion, economics, and law have addressed work and nonwork factors from specialized perspectives, but today, changes in the world of work require a holistic view. There are numerous definitions of well-being within

* Reprinted verbatim with permission from Schulte, Paul A. et al. (2015), Considerations for incorporating "well-being" in public policy for workers and workplaces, *American Journal of Public Health*, Vol. 105, No. 8, pp. e31–e44, August 2015. doi:10.2105/AJPH.2015.302616 (Authors: Paul A. Schulte, Rebecca J. Guerin, Anita L. Schill, Anasua Bhattacharya, Thomas R. Cunningham, Sudha P. Pandalai, Donald Eggerth, and Carol M. Stephenson.)

and between disciplines, with subjective and objective orientations addressing such conceptualizations as happiness, flourishing, income, health, autonomy, and capability (Ringen, 1996; Danna and Griffin 1999; Cronin de Chavez, 2005; Adler, 2011; Diener, 2013; Fleuret, 2007; Ng ECW, 2013; Seligman, 2011; Ryff, 2014; Agrawal, 2001; Warr, 1987). Well-being is widely cited in public policy pronouncements, but often in the conjunctive form of "… and well-being" (as in health and well-being). It is rarely defined or operationalized in policy. In this chapter, we consider if the concept of "well-being" is useful in addressing contemporary issues related to work and the workforce and, if so, whether it can be operationalized for public policy and what the implications are of doing so. We discuss the need to evaluate a broad range of work and nonwork variables related to worker health and safety and to develop a unified approach to this evaluation. We discuss the potential of well-being to serve as a unifying concept, with focus on the definitions and determinants of well-being. Within this part of the discussion, we touch on topics of responsibility for well-being. We also explore issues of importance when one is incorporating well-being into public policy. We present examples of the incorporation of the principles of well-being into public policy, and the results thus far of the implementation of such guidance. We describe research needs for assessing well-being, particularly the need to operationalize this construct for empirical analysis. We aim to contribute to the ongoing efforts of occupational safety and health and public health researchers, practitioners, and policymakers to protect working populations.

New patterns of hazards

Many of the most prevalent and significant health-related conditions in workers are not caused solely by workplace hazards, but also result from a combination of work and nonwork factors (including such factors as genetics, age, gender, chronic disease, obesity, smoking, alcohol use, and prescription drug use). (Schulte et al., 2012; Pandalai et al., 2013) One manifestation of this interaction of work and nonwork risks is the phenomenon known as presenteeism—diminished performance at work because of the presence of disease or a lack of engagement—which appears to be the single largest cause of reduced workforce productivity (Hemp, 2004; Bustillos, 2013; de Perio et al., 2014; Johns, 2010). Chronic disease is also a major factor in worker, enterprise, and national productivity and well-being. The direct and indirect costs of chronic disease exceed $1 trillion annually. (DeVol et al., 2007) Moreover, the costs of forgone economic opportunity from chronic disease are predicted to reach close to $6 trillion in the United States by 2050. (DeVol et al., 2007) Chronic disease places a large burden on employers as well as on workers.

Although the classic significant occupational hazards still remain, work is changing. Many transitions characterize the twenty-first-century economy: from physical to more mental production, from manufacturing to service and health care. New ways of organizing—contracting, downsizing, restructuring, lean manufacturing, contingent work, and more self-employment (Stone, 2004; Howard, 2010; Holzer, 2007; Cummings, 2008; Ng ECW, 2013; Tehrani et al., 2007; Osterman and Shulman, 2011)—affect many in the labor force, and heightened job uncertainty, unemployment, and underemployment also characterize work in the global marketplace. (Stone, 2004; Howard, 2010; Kitt 2013; Sparks, Faragher, Cooper et al., 2001; Guillemin, 2011; Osterman and Shulman, 2011)

Increasingly, work is done from home. Sedentary work accounts for larger portions of time spent in the workplace (Church, Thomas, Tudor-Locke et al., 2011). New employment relations and heightened worker responsibilities are manifest in what organizational theorists have termed the "psychological contract" to characterize the nature of employers' and employees' mutual expectations and perceptions of work production (Stone, 2004). It may be as Stone has concluded, that "the very concept of workplace as a place and the concept of employment as involving an employer are becoming outdated in some sectors" (Stone, 2004: p. ix). The demographics of the workforce are also changing. More immigrants and women are joining the labor force, and older workers (those aged 55 years and older) are now the largest and fastest-growing segment of the workforce. (Costa et al., 2005; Shrestha, 2006; Flynn, 2014)

Immigrants are projected to make up roughly 23% of working-age adults by 2050, (Flynn, 2014; Passel JS, Cohn, 2008; Cierpich et al., 1992–2006) a demographic shift that has significant implications on and for public health. Among other issues, Latino immigrants suffer significantly higher workplace mortality rates (5.0 per 100,000) than all workers (4.0 per 100,000) (Flynn, 2014; Passel JS, Cohn, 2008; Cierpich et al., 1992–2006). There is growing participation of men and women in previously gender-segregated fields and the integration of 2.4 million veterans who have served in Iraq and Afghanistan since 2011 (Flynn, 2014; Passel JS, Cohn, 2008; Cierpich et al., 1992–2006; Waterstone, 2010).

In some countries, workers are voluntarily working later in life; in others, decreased dependency ratios require them to work longer (Shrestha, 2006; Rissa and Kaustia, 2007; Harter et al., 2003). There also are many workers wanting to work but without the necessary skills to meet job requirements (Kochan, 2012).

The unifying concept of well-being

With the seismic shifts occurring in the global economy, there is a need for a concept that unifies all the factors that affect the health of workers, as well as the quality of their working lives (Antonovsky, 1996; Dewe and Kompier, 2008; Bauer, 2010; Schulte, 2010; Vainio, 2012). Well-being may be such a concept and, furthermore, it could be addressed as either an outcome in and of itself that might be improved by the application of prevention or intervention strategies that target risk factors, or it may be looked at as a factor that influences other outcomes. For example, healthy well-being has been linked to such outcomes as increased productivity, lower health care costs, health, and healthy aging (Rissa and Kaustia, 2007; Harter et al., 2003; Prochaska, 2011; Budd and Spencer, 2014; Gandy et al., 2014; Grawitch et al., 2006). Numerous studies have shown links between individuals' negative psychological well-being and health (Huppert, 2009; Chida and Steptoe; 2008). Conversely, evidence from both longitudinal and experimental studies demonstrates the beneficial impact of a positive emotional state on physical health and survival (Pressman et al., 2015, Diener and Chan, 2011, Howell et al., 2007). In these contexts, well-being might be thought of as a factor that has an impact on various outcomes. In other contexts, well-being is the outcome. Studies have shown that working conditions, including such factors as stress, respect, work–life balance, and income, affect well-being (Noor, 2003; Karanika-Murray and Weyman, 2013; Hodson, 1999, 2001; Mellor and Webster, 2013).

Well-being has been defined and operationalized in different ways, consistent with roles either as a risk factor or as an outcome. Depending on the context, the literature associating well-being with productivity, health care costs, and healthy aging is heterogeneous, of variable quality, and difficult to assess and compare. Nonetheless, there are well-designed studies in different disciplines that indicate the potential benefits of using and operationalizing the concept of well-being (Ringen, 1999; Adler, 2011; Costa et al., 2005; Prochaska et al., 2011; Huppert, 2009; Boehm et al., 2011; Warr, 1990; Linley, 2009).

The use of an overarching concept such as "well-being" in public policy may more accurately capture health-related issues that have shifted over the past 30 years along with changes in population demographics, disease patterns, and the world of work. These shifts, which include changes in the nature of work, the workforce, the workplace, and the growing impact of chronic disease, have resulted in complex issues that interact and do not fit neatly in one field or discipline, making them especially difficult to address (Holzer and Nightingale, 2007; Sparks et al., 2001; Cronin de Chavez, 2005; Ng ECW, 2013).

Definitions, determinants, and responsibility for well-being

The published literature abounds with many definitions of well-being. These definitions may be objective, subjective (with a focus on satisfaction, happiness, flourishing, thriving, engagement, and self-fulfillment) or some synthesis of both (Danna, 1999; Cronin de Chavez, 2005; Prochaska

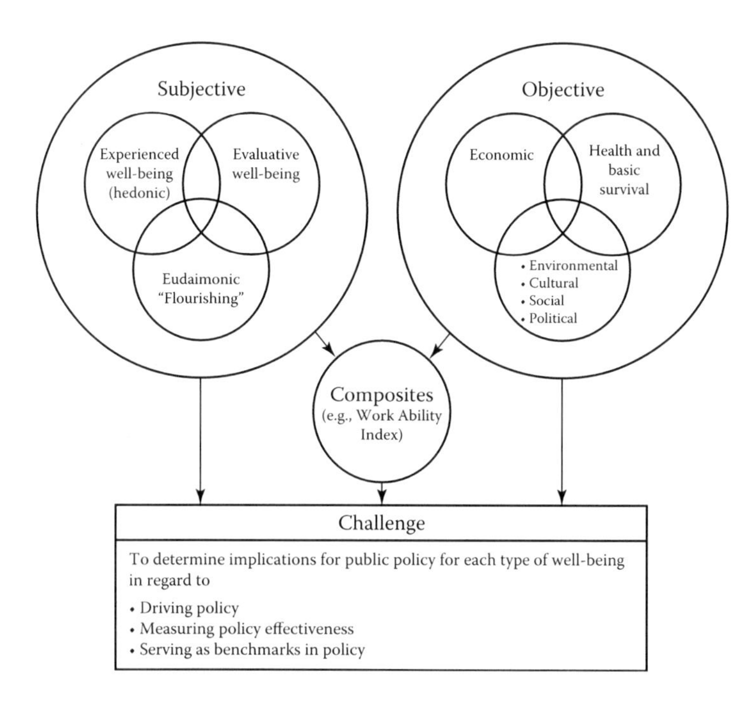

Figure 11.1 Types of well-being and policy challenges. (Based on National Research Council, Paper on measuring subjective well-being in a policy framework, Washington, DC, National Academies Press, 2013; Xing Z and Chu L, Research on constructing composite indicator of objective well-being from China mainland, *Proceedings of the 58th World Statistical Congress*, Dublin, Ireland, International Statistical Institute, 2011, 3081–3097.)

et al., 2011; Kahneman et al., 1993, Sen, 1993; Atkinson et al., 2012; Adler, 2013; National Research Council, 2013; Dolan and White, 2007). With regard to work, some definitions focus on the state of the individual worker, whereas others focus on working conditions, and some focus on life conditions. Figure 11.1 shows three major types of well-being—objective, subjective, and composite—and the policy challenges related to them.

Objective well-being

Defining and measuring well-being objectively for use in policy is to be able to make well-being judgments that are not purely subjective, that are independent of preferences and feelings, and that have some social agreement that such measures are basic components of well-being (Gaspart, 1997). Thus, having enough food, clothing, and shelter are rudimentary components of objective well-being. Their absence correlates with lack of well-being. For workers, income, job opportunity, participation, and employment are examples of components or facets of objective well-being. For companies, absenteeism, presenteeism, workers' compensation claims, and productivity are objective indicators of well-being. The importance of measuring well-being with an objective index has been expressed in philosophical debates on distributive justice (Rawls, 2009; Sen, 1999; Cohen, 1993).

If well-being is measured objectively, its index applies to everybody in the workforce so levels of well-being can be compared (National Research Council, 2013). However, when objective well-being is assessed by using "objective lists," there are questions of what should be on the list and how it should be ordered (Dolan, 2007).

Objective well-being has various domains. Xing and Chu identified six major domains of objective well-being: health and basic survival, economic, environmental, cultural, social, and political (Figure 11.1) (Xing and Chu, 2011).

In recent years, the use of objective measures of well-being has been questioned and a call has been made for more overarching indicators of well-being (Otoiu et al., 2014; Sen, 1999; Stiglitz et al., 2009). This is in part attributable to the fact that objective well-being indicators, such as economic ones, do not always adequately portray the quality of life. In 1974, Easterlin demonstrated that increases in income do not match increases in subjective well-being (Easterlin, 1974). This is because people's aspirations change in line with changes in their objective circumstances (Dolan and White, 2007).

Subjective well-being

Subjective well-being is multifaceted and includes evaluative and experienced well-being (hedonic) and eudaimonic well-being, which refers to a person's perceptions of meaningfulness, sense of purpose, and value of his or her life (National Research Council, 2013). Subjective well-being as originated by Warr has been described as assessable on three axes: displeasure-to-pleasure, anxiety-to-comfort, and depression-to-enthusiasm (Warr, 1999). Research has shown that subjective well-being measures relate in a predictable manner to physiological measures, such as cortisol levels and resistance to infection (Steptoe et al., 2005). More specifically for the work environment, well-being can be described in a model that identifies job-specific well-being that includes people's feelings about themselves, in their job, and more general feelings about one's life, referred to as "context-free" well-being (Warr, 1999; Hassan et al., 2009).

If subjective well-being is to be used in policy, the measures of it have to be shown as valid, reproducible, and able to indicate "interpersonal cardinality" because individuals may interpret scales differently.

Interpersonal cardinality means that when two persons rate their subjective well-being as, for example, "5 out of 10," they mean the same thing (Bronsteen et al., 2014). Consideration of subjective well-being raises the question of the extent that it is related to personality traits, emotions, psychological traits that promote positive emotions, and the relationship with health status (Ryan and Deci, 2001). Subjective well-being also needs to be considered with regard to whether it changes across the life span and with aging. The main challenge with measuring subjective well-being is the different conclusions that can be drawn depending on the numbers of factors that are accounted for and then controlled for in the analysis (Buffet et al., 2013). When considered for the workforce or workplace policy, the question arises, should well-being at work be separated from well-being generally? Moreover, if well-being is included in policy, how do countries and organizations move from policy to practice? (Buffet et al., 2013).

In discussions of subjective well-being in relation to policy, the focus is often on the measurement of experienced well-being. The unique policy value of experienced well-being measures may not be in discovering how clearly quantifiable factors (such as income) relate to aggregate-level emotional states, but rather in uncovering relationships that would otherwise not be acted upon (National Research Council, 2013). For example, levels of activity or time use in domains such as commuting or exercise may have an impact on well-being in ways that are not easily captured by objective measures. Subjective well-being measures seem most relevant and useful for policies that involve assessment of costs and benefits when there are not easily quantifiable elements involved—for instance, government consideration of spending to redirect an airport flight path to reduce noise pollution, funding alternative medical care treatments when more is at stake than maximizing life expectancy, or selecting between alternative recreational and other uses of environmental resources (National Research Council, 2013).

Thus far, evidence about interactions between experienced well-being and other indicators is inconclusive. For example, on the relationship between income and experienced well-being, Deaton and Stone noted that, at least cross-nationally, the relationship between aggregate positive emotions (here, meaning day-to-day experienced well-being) and per capita gross domestic product is unclear (Deaton and Gallus, 2013).

Some authors have suggested that an important consideration in using subjective well-being as a national aggregate indicator of well-being is that it may then easily become manipulated in the reporting of well-being (Frey, 2013). Thus, it is suggested that there may be an inherent limitation in selecting subjective well-being as a target of policy in that its measurement may become unreliable or biased. Ultimately, subjective well-being has not been recommended as a singular indicator of social well-being but rather as an additional indictor that can capture a range of factors that are difficult to quantify, which may have an impact on well-being (National Research Council, 2013). Thus, a subjective assessment of well-being should serve as a supplement to other key social statistics.

Composite well-being

Composite indicators of well-being combine objective and subjective indicators of well-being into a single measure or measures. The underlying premise of composite well-being is that both objective and subjective well-being are important, complementary, and needed in public policy development, implementation, and assessment. There are various examples of composite well-being. The Gallup–Healthways Index is a composite measure composed of five domains: life evaluation, emotional health, work environment, physical health, and basic access (Harter et al., 2003; Rath et al., 2010). Some of the domains are subjective, some are objective. More specific for work, the Work Ability Index (WAI) developed in the 1980s in Finland has been widely assessed and can be seen as a composite indicator of well-being. Its validity has been determined and it has been evaluated in numerous populations.

Workability indicates how well a worker is in the present and is likely to be in the near future, and how able he or she is to do his or her work with respect to work demands, health, and mental resources (Ilmarinen et al., 1991). Use of the WAI in the Netherlands is an example of using composite well-being in policy. The Dutch Ministry of Social Affairs established in 2008 a program (Implementation of WAI in the Netherlands) to promote, understand, and use workability (ESF Age Network, 2015).

A database of the use of the WAI and related job training, job designs, health, and career interventions to improve WAI is used to benchmark among sectors and is then used by policymakers to identify where more intervention is needed. Composite indicators of well-being in general are subject to the impact of the selection of indicators, high correlation between components, computational focus, and component weighting (Otoiu et al., 2014).

A variant of the composite view of well-being is the simultaneous measurement of multiple aspects or domains of well-being (Smith and Clay, 2010). For example, a program designed to improve well-being in a retail workforce resulted in improvement of multiple domains of well-being measured independently as improved productivity and increased overall profitability (Rajaratnam et al., 2014). In another study, Smith and Clay showed that plotting at least one subjective and one objective measure of well-being can be used to compare activities at multiple scales and across nations, time, and cultures (Smith and Clay, 2010).

Definitional clarity and measurement

There has been a marked tendency to think of well-being as a synonym for health or for mental health. However, broader definitions have been widely used. In addition to objective, subjective, and composite well-being definitions, well-being has been examined as either a risk factor for physical and mental health outcomes, or as an outcome impacted by personal, occupational, or environmental risk factors, or both at the same time. The definitions and resultant policy pronouncements involving well-being involve very diverse conceptualizations in terms of scale, scope, location, and responsibility (Cronin de Chavez et al., 2005; Fleuret and Atkinson, 2007; Hodson, 1999; Boehm et al., 2011; Rath et al., 2014). Although a single definition of well-being may not be needed, there is need for clarity and specification of the ones used to drive policy, as a component of policy, or to evaluate policy (Adler, 2013; Buffet et al., 2013; Allin, 2014). There have been different focal formulations of well-being such as well-being at work, well-being of workers, employee well-being, workforce well-being, workplace well-being, and well-being through work. In some cases, the difference in focus may be merely semantics. However, the focus of well-being may influence determinants, measures, variable selections, interventions, and ultimately regulation and guidance. For "well-being" to be a useful concept in public policy, it has to be defined and operationalized so that its social and economic determinants can be identified and interventions can be developed so that inducements to "ill-being" in the activities and relations in which people participate can be addressed (Neubauer and Pratt, 1981).

Various tools, instruments, and approaches (e.g., questionnaires, databases, and economic and vital statistics) have been used to measure well-being. The type of measurement may depend on the type of well-being (i.e., objective, subjective, composite) and level of consideration (i.e., individual worker, enterprise, or national level). Warr, 2012 identified eight elements that should be addressed in measurement of well-being:

1. Whether it should be seen from a psychological, physiological, or social perspective
2. Whether it should be viewed as a state (time specific) or a trait (more enduring)
3. Its scope (i.e., type of setting or range)
4. Whether to focus on the positive or negative aspects of well-being or a combination of both
5. Viewing them as indicators of affective well-being and cognitive: Affective syndromes (i.e., considering feelings only or including perceptions and recollections)
6. What to assess when one is measuring affective well-being

7. What to assess when one is measuring syndrome well-being
8. Examining ambivalence, which includes the temporal aspects, at one point in time or across time (Buffet et al., 2013)

Determinants of well-being

The literature on determinants of well-being of the workforce as an outcome attributable to work-related factors is extensive and diverse (Danna and Griffin, 1999; Prochaska et al., 2011; Karanika-Murray and Weyman, 2013; Hodson, 1999; Warr, 1999; Buffet et al., 2013). A large number of factors have been investigated and implicated (Danna and Griffin, 1999; Karanika-Murray and Weyman, 2013; Mellor and Webster, 2013; Rounds et al., 1987; Karasek, 1979; Karasek, 1992; Grebner et al., 2005; Johnson and Hall, 1988; Dhondt et al., 2014; Bakker and Demerouti, 2007; Siegrist, 1996). Chief among these are workplace management, employee job control, psychological job demands, work organization, effort and reward, person–environment fit, occupational safety and health, management of ill health, and work–life balance.

Using a longitudinal design and external rating of job conditions, Grebner et al. observed that job control (i.e., feelings of control over one's work) correlated with well-being on the job, and job stressors correlated with lower well-being (Grebner et al., 2005). Hodson identified the main structural determinants of worker well-being through a quantitative analysis of ethnographic accounts of 108 book-length descriptions of contemporary workplaces and occupations (Hodson, 1999, 2001). Narrative accounts were numerically coded and analyzed through a multivariate analysis (Hodson, 1999, 2001). The research determined that the strongest determinants of worker well-being were mismanagement, worker resistance (any individual or small group act to mitigate claims by management on employees or to advance employees' claims against management), and citizenship (positive actions on the part of employees to improve productivity and cohesion beyond organization requirements) (Hodson, 1999, 2001). A large amount of literature has quantified job demands and control as they pertain to workers' well-being defined by job stress (Boehm et al., 2011; Karasek, 1979, 1992; Grebner et al., 2005; Johnson and Hall, 1988; Dhondt et al., 2014; Bakker and Demerouti, 2007; Siegrist, 1996).

Well-being of workers has also been assessed by using the effort–reward imbalance model, which addresses the effort workers put into their jobs and the rewards they get. When the rewards comport with the efforts, well-being is increased (Siegrist, 1996). In 1979, Karasek suggested that jobs be redesigned to include well-being as a goal (Karasek, 1979). The relationship of well-being in terms of health has been assessed with regard to job insecurity, work hours, control at work, and managerial style (Sparks, 2001; Allin, 2014; Warr, 2012; Rounds et al., 1987). Well-being at work has also been assessed in terms of factors such as work–life balance, wages, genes, personality, dignity, and opportunity (Noor, 2013; Karanika-Murray and Weyman, 2013; Hodson, 1999, 2001; Warr, 1999; Bockerman et al., 2011; Clark, 1996; Blanchflower and Oswald, 2004; Frey and Stutzer, 2002; Diener et al., 2003).

Another approach used in vocational psychology and job counseling to assess determinants of well-being is the use of "person–environment" fit models. These models make the simple prediction that the quality of outcomes directly reflects the degree to which the individual and the environment satisfy the other's needs. The Theory of Work Adjustment has detailed a list of 20 needs common, at varying degrees, to most individuals and most workplaces (Dawis and Lofquist, 1984; Blustein, 2006). It also outlines the mechanics and dynamics of this interaction between person and environment to make predictions about outcomes and it is one of the few models that give equal weight to satisfaction of the worker and the workplace (most approaches emphasize one at the expense of the other—or even ignore one entirely).

In yet another approach, macroergonomics evolved to expand on the traditional ergonomics of workstations and tool design, workspace arrangements, and physical environments (Punnett et al., 2013; Hendrick and Kleiner, 2005). Macroergonomics addresses the work organization, organizational structures, policies, climate, and culture as upstream contributors to both physical and mental health. This approach helps to translate concepts of well-being to specific job sites (Blustein, 2006; Punnett et al., 2013). Ultimately, if the well-being of the workforce is to be achieved and maintained, the policies and guidance need to be practical and feasible at the job level (National Institute for Occupational Safety and Health, 2010). The well-being of the workforce is dependent not only on well-being in the workplace, but also on the nonwork determinants of well-being. These determinants include personal risk factors (e.g., genetics, lifestyle) as well as social, economic, political, and cultural factors.

The challenge in accounting for these may not be to use every factor but to select factors that are surrogates for or represent constellations of related determinants.

Responsibility for well-being

Well-being is a summative characteristic of a worker or workforce and it can also be tied to place, such as to the attributes of the workplace (Fleuret and Atkinson, 2007; Schulte and Vainio, 2010; Ryan and Deci, 2001). In part, achieving increased well-being or a desired level of well-being of the worker and the workforce is inherent in the responsibility of the employers to provide safe and healthful work. However, because well-being is a summative concept that includes threats and promoters of it, as well as nonwork factors, the worker (employee) has a responsibility, too. Clearly, this overlapping responsibility is a slippery slope that could lead to blaming the worker for decreased well-being. Thus, it may be necessary to distinguish work-related from non-work-related sources of well-being and to identify the apportionment of responsibility. This is a new and challenging endeavor. The standard practice in occupational safety and health is that the employer is responsible for a safe and healthy workplace and the employee is responsible for following appropriate rules and practices that the employer establishes to achieve a safe and healthy workplace.

If a higher-level conceptualization of well-being is pursued, which subsumes health, is aspirational, and includes reaching human potential, then workers surely must actively engage in the process. How the roles of employer versus employee will be distinguished, and what to do about areas of overlap, are critical questions that need to be addressed. In addition, although well-being at work may be primarily an employer's responsibility, well-being of the worker or workforce is also the responsibility (or at least in the purview) of others in society (e.g., governments, insurance companies, unions, faith-based and nonprofit organizations) or may be affected by nonwork domains. Clearly, the well-being of the workforce extends beyond the workplace, and public policy should consider social, economic, and political contexts.

Issues for including well-being in public policy

Two critical policy considerations address the incorporation of well-being for workers in public policy issues. First is the question of how well-being will be used, and second is what actions should be taken to reduce threats to well-being and increase promotion of it. The initial need is to consider the question of how the concept will be used. Will well-being be an end (dependent variable), a means to an end (independent variable), or an overarching philosophy? The dominant observation of this chapter is that well-being in policy has been used as an overarching philosophy. The 1998 Belgian Legislation on the policy of well-being of workers at work is such an example. It pertains to

1. Work safety
2. Health protection of the worker at work
3. Psychosocial stress caused by work including violence, bullying, and sexual harassment
4. Ergonomics
5. Occupational hygiene
6. Establishment of the workplace, undertaking measures relating to the natural environment in respect to their influence on points 1 through 6 (Belgian Legislation, 2015; Anttonen and Räsänen, 2008).

For the most part, this decree mandates what is considered as the current practice of occupational safety and health. This example does not capture some components of well-being such as self-fulfillment, engagement, flourishing, and opportunity often found in many definitions. Even if it did, this use as an overarching philosophy may serve to drive policy but does not necessarily identify or address determinants or measures of whether well-being is achieved. If well-being is considered as a policy objective, then its presence or degree needs to be measurable. Are there attributes of well-being that could be defined and ultimately measured? Numerous instruments for measuring well-being in general and some for well-being in the workplace have been developed (Warr, 1987; Harter et al., 2003; Prochaska et al., 2011; National Research Council, 2013; Ilmarinen et al., 1991; Anttonen and Räsänen, 2008; European Working Conditions Observatory, 2008). Sometimes the measurements are surrogates or components of well-being (such as longer careers, health, or productivity); other times they involve the capacity to attain well-being (such as income, job control, and autonomy) (Sen, 1993; Diaz-Serrano and Cabral Vieira, 2005; Anttonen, 2009; Michaelson et al., 2009; Pot et al., 1994; Osberg, 2004).

In terms of work, well-being could be described in terms of the worker overall, at work and outside work. It can be considered in terms of the workforce as a whole, by sector, by enterprise, and by geographic designation. The consideration of well-being at work could be addressed in public policy through performance or specification approaches. The "performance" approach would stipulate an end state, possibly a state of well-being demonstrated by positive indicators (e.g., worker engagement, decreased absenteeism, increased productivity, decreased bullying, or harassment); however, the means to that end would not be specified. By contrast, in a "specification" approach, the means to achieve an end would be specified.

Actions to reduce threats to, and increase promotion of, well-being

The second critical policy question is what actions should be taken to reduce threats to well-being of workers and increase promotion of it. This will in part need to be addressed in terms of where regulatory authority circumscribes the actions that must be taken to protect workers and prevent workplace disease, injury, and deaths. Therefore, in the United States, the focus would be on the Occupational Safety and Health Act, the Mine Safety and Health Act, the Toxic Substances Control Act, and other legislation that has worker safety and health provisions.

The definition of health in work-related legislation has generally been narrowly originated to be the absence of disease. If, however, a broader definition of health, such as the one promoted by World Health Organization (WHO) (WHO, 1948) is used, well-being is a central component. The WHO definition is "Health is a state of complete physical, mental and social well-being and not merely the absence of disease or infirmity"(WHO. 1948, p. 1). As a consequence, using this definition of health alters the interpretation of legislation dealing with workers' health. However, one of the unintended interpretations of the WHO definition is that, in the era of chronic disease, a goal of complete well-being could lead to the "medicalization of society" where many people would be considered unhealthy most of the time because they lack complete well-being. If the WHO definition is reformulated to a more dynamic view, "based on the resilience or capacity to cope and maintain one's integrity, equilibrium and sense of well-being," it would better be a focus for public policy (Huber et al. 2011, p. 2).

In addition, in some of the current legislation, there is the implication to go beyond health and address broader issues of well-being.

For example, in the United States, the 1970 Occupational Safety and Health Act (Pub. L. No. 91–596, 84 STAT. 1590, December 29, 1970) also mandates the preservation of human resources, although interpreting this statement broadly to include "well-being" was probably not the original legislative intention; its inclusion, however, is consistent with considering a range of factors influencing workers' well-being. Although this extant legislation could be sufficient, it may be that additional legislation or consensus standards are needed that will identify responsibilities for achieving or maintaining workforce well-being.

Well-being as a driver of public policy

In addition to being included in public policies, well-being assessments may be used to affect or drive public policy (Adler, 2013; Allin, 2014; Atkinson et al., 2011). They may serve as part of a feedback loop on existing conditions or policies to define success, failure, or the need for modification of them (Allin, 2014; Huppert et al. 2009). Because worker well-being is a multifactorial concept, it may be important to understand which aspect of well-being is affected by specific policies or conditions (Ng and Fisher, 2013; Rajaratnam et al., 2014; Allin, 2014). For example, one of the best surveys related to well-being of workers is the European Working Conditions Survey, which was first conducted in 1990 and periodically repeated in 16 countries (European Working Conditions Observatory, 2008). This household survey addresses a broad range of themes, which taken together can be seen as describing well-being in a broad sense of the term. Many of the individual themes of the survey involve various components of well-being and are strong predictors of well-being (European Working Conditions Observatory, 2008; Diaz-Serrano and Cabral Vieira, 2005). These themes include employment status, working time duration, safety, work organization, work–life balance, worker participation, earnings, and financial security, as well as work and health. The results of this survey are periodically compared for 16 European Nations. Use of the survey in the United States would be a useful way to identify a baseline on well-being that could serve to drive public policies as well as serve as a comparison over time. Although this survey is a useful tool, it is subjective and only addresses well-being from the worker perspective. There is also need for tools to assess well-being from the societal perspective.

Well-being can also be used to screen public policies. Screening can assess whether a policy will have an effect on workforce well-being or its drivers (Allin, 2014). It is also worth reflecting on at what stage in policy development and appraisal it is best to consider well-being. Allin identified stages of a broad policy appraisal cycle and incorporated a well-being perspective (Allin, 2014). Figure 11.2 traces the steps for developing, implementing, and evaluating policy. Well-being can be considered at each step in various ways.

Examples of the use of well-being in public policy or guidance

Various efforts illustrate application of the well-being concept in policies, guidance, and research (Table 11.1). The WHO document "Healthy workplaces: A model for action: For employers, workers, policy-makers, and practitioners" uses the concept of well-being as a foundation (Burton, 2010). Building on the WHO definition of health (WHO, 1948), the document defines a healthy workplace as one in which workers and managers collaborate to use a continual improvement process (based on the work of Deming [Deming, 2008]) to protect the health, safety, and well-being of all workers and the sustainability of the workplace. Four areas identified for actions that contribute to a healthy workplace include the physical work environment, the psychosocial work environment, personal health resources in the workplace, and enterprise community involvement.

The National Institute for Occupational Safety and Health Total Worker Health program integrates occupational safety and health protection with workplace policies, programs, and practices that promote health and prevent disease to advance worker safety, health, and well-being (Schill and Chosewood, 2013; Howard, 2013). It thus explicitly focuses on both the workplace and the worker as well as on the dynamics of employment. There is a growing body of support for prevention strategies to combine health protection and health promotion in the workplace (National Institute for Occupational Safety and Health, 2010; Schill and Chosewood, 2013; Howard, 2013; Hammer and Sauter, 2013). The Total Worker Health program is an expanded view of human capital, identifying what affects a worker as a whole rather than distinguishing between "occupational" and "nonoccupational realms" (Howard, 2013; Hammer and Sauter,

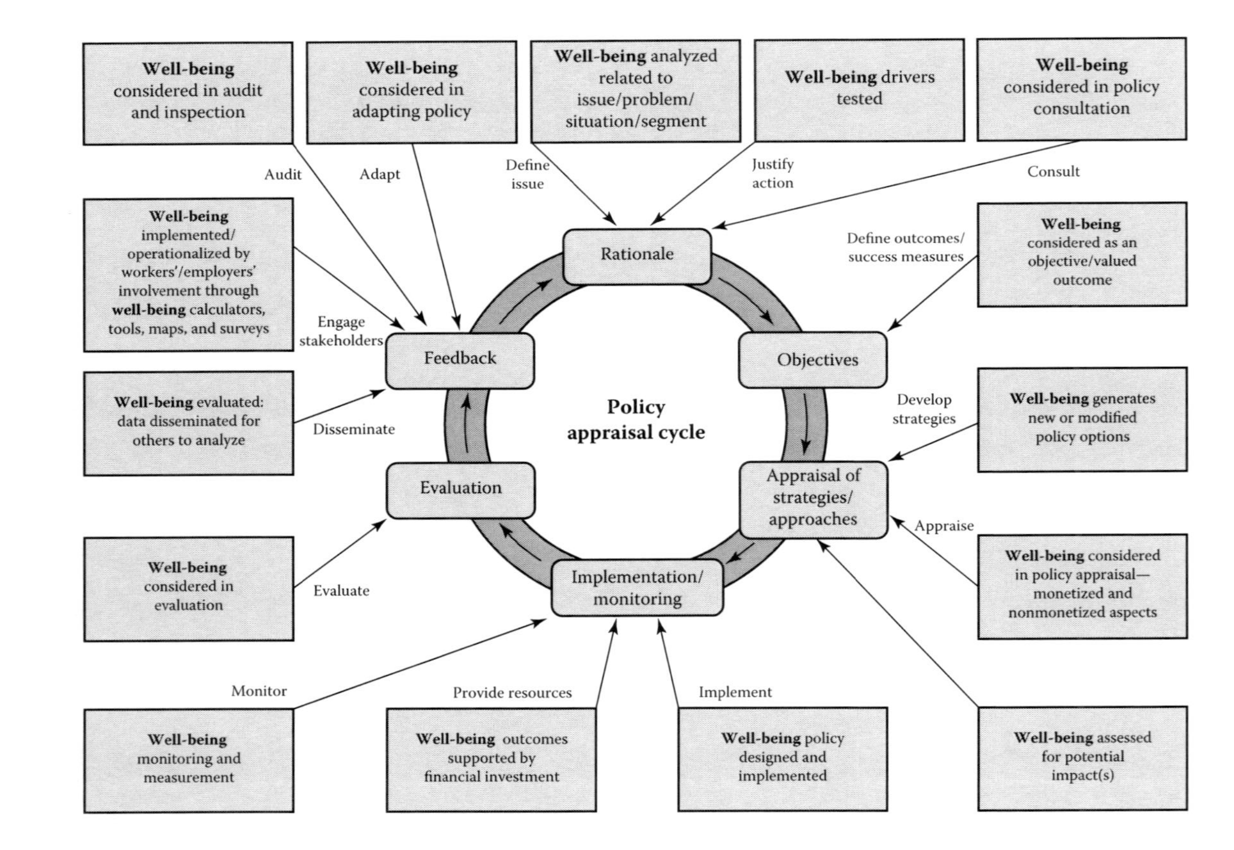

Figure 11.2 Worker/workforce well-being and the policy appraisal cycle. (Adapted from Allin P, *Work and Wellbeing: A Complete Reference Guide*, West Sussex, UK, John Wiley, 2014.)

Table 11.1 Example of well-being in research, guidance, and regulation

References (Year)	Use	Conception or definition of well-being	Focus or methods	Metrics or indications
Research				
Harter et al. (2003)	Relationship to business outcomes	"We see well-being as a broad category that encompasses a number of workplace factors. Within the overall category of well-being we discuss a hypothesized model that employee engagement (a combination of cognitive and emotional antecedent variables in the workplace) generates higher frequency of positive affect (job satisfaction, commitment, joy, fulfillment, interest, caring). Positive affect then relates to the efficient application of work, employee retention, creativity, and ultimately business outcomes."(p.2–3)	Meta-analysis of 36 companies involving 198,514 respondents	Employee engagement-linked business outcomes
Chida and Steptoe (2008)	Relationship to mortality	"Positive psychological well-being encompasses positive affect and related trait-like constructs or dispositions, such as optimism.... Positive affect can be defined as a state of pleasurable engagement with the environment eliciting feelings, such as happiness, joy, excitement, enthusiasm, and contentment."(p.741)	Systematic review of prospective cohort studies to determine link between well-being and mortality	35 studies, hazard ratio = 0.82; 95% confidence interval = 0.76, 0.89

(*Continued*)

Table 11.1 (Continued) Example of well-being in research, guidance, and regulation

References (Year)	Use	Conception or definition of well-being	Focus or methods	Metrics or indications
Huppert et al. (2009)	European Social Survey	"Well-being is a complex construct."(p.303) It is "people's perceived quality of life, which we refer to as their 'well-being.'"(p.302)	Well-being module and questionnaire	54 items, divided into 2 sections, corresponding to personal and interpersonal dimensions of well-being; each of these is further subdivided into feeling (being) and functioning (doing)
Guidance or practice, Delwart et al. (2011)	Well-being in the workplace guide		Identifying best practices related to well-being in the workplace	Key performance indicators
Schill and Choosewood (2013)	Guidance for Total Worker Health	"Total Worker Health™ is a strategy integrating occupational safety and health protection with health promotion to prevent worker injury and illness and to advance health and well-being." (p.58) "Promoting optimal well-being is a multifaceted endeavor that includes employee engagement and support for the development of healthier behaviors, such as improved nutrition, tobacco use/cessation, increase physical activity, and improved work/life balance."(p.59)	Guidance on how to integrate health protection and promotion	Various score cards

(Continued)

Table 11.1 (Continued) Example of well-being in research, guidance, and regulation

References (Year)	Use	Conception or definition of well-being	Focus or methods	Metrics or indications
Canadian Standard Association (2013)	Mental well-being guidelines	Psychological health and safety in the workplace.	Developing an action plan for implementing the mental health and well-being strategy	Integrated approach
Burton (2010)	WHO model for healthy workplaces	"WHO's definition of health is: 'A state of complete physical, mental and social well-being and not merely the absence of disease.' In line with this, the definition of a healthy workplace … is as follows: A healthy workplace is one in which workers and managers collaborate to use a continual improvement process to protect and promote the health, safety, and well-being of all workers and the sustainability of the workplace."(p.90)	Framework for health protection and promotion	Continuous improvement method
Regulation or policy, Pot et al. (1994)	WEBA (conditions of well-being at work policy)	Operational well-being in terms of organization of work and ergonomics.	How to measure individual well-being	3 value levels for 7 questions
East Riding of Yorkshire Council (2004)	Psychological well-being at work policy	The mental health as well as the physical health, safety, and welfare of employees.	Stress assessment (relating to work pressures) tool to develop action plan	Likert-like scoring; risk scoring

(Continued)

Table 11.1 (Continued) Example of well-being in research, guidance, and regulation

References (Year)	Use	Conception or definition of well-being	Focus or methods	Metrics or indications
Ministry of Social Affairs and Health (2011)	Well-being at work policies	"Health, safety and well-being are important common values, which are put into practice in every workplace and for every employee. The activities of a workplace are guided by a common idea of good work and a good workplace. Good work means a fair treatment of employees, adoption of common values as well as mutual trust, genuine cooperation and equality in the workplace. A good workplace is productive and profitable."(p.4) "From the perspective of the work environment, a good workplace is a healthy, safe and pleasant place. Good management and leadership, meaningful and interesting tasks, and a successful reconciliation of work and private life are also characteristics of a good workplace."(p.4)	Policies to specify a ministerial strategy	Extend work 3 years Reduce workplace accidents 25%; reduce psychic strain 20%; reduce physical strain 20%

Note: WEBA = conditions of well-being at work; WHO = World Health Organization.

2013). The program calls for a comprehensive approach to worker safety, health, and well-being. Guidelines and frameworks are provided to implement policies and programs that integrate occupational safety and health protection with efforts to promote health and prevent disease. In another effort, the recent Canadian consensus standard on mental health in workers illustrates a step in the direction of targeting psychological well-being as an outcome, and the means to achieve it (Canadian Standards Association, Bureau de normalization du Quebec, 2013). The stated purpose of this voluntary standard is to provide a systematic approach for creation of workplaces that actively protect the psychological health and promote the psychological well-being of workers. "Psychological well-being" is characterized by a state in which the individual realizes his or her own abilities, can cope with the normal stresses of life, can work productively and fruitfully, and is able to make a contribution to his or her community (Canadian Standards Association, Bureau de normalization du Quebec, 2013). To achieve this purpose, the standard details requirements for development, implementation, evaluation, and management of a workplace psychological health and safety management system (PHSMS). The standard enumerates 13 workplace factors that affect psychological health and safety and uses these factors to guide the conceptualization of a PHSMS. Risk management, cost-effectiveness, recruitment and retention, and organizational excellence and sustainability are included in the business case rationale for a comprehensive and effective PHSMS. In addition to an outcome of improved well-being, benefits for workers from a PHSMS include improved job satisfaction, self-esteem, and job fulfillment (Canadian Standards Association, Bureau de normalization du Quebec, 2013). Additional examples regarding the use of the term "well-being" can be found in a number of European policies (Anttonen and Räsänen, 2008). In the Netherlands, guidance was issued in 1989 that called for well-being in the workplace. This legislation operationalized well-being in two categories: organization of work and ergonomics. It also required the training of a "new" type of professional at the master's degree level with expertise in organization of work (Pot et al., 1994). However, after a few years, the legislation was challenged because of difficulty in enforcing the subjective concept of well-being, and nullified by the courts. During that period, an instrument to assess job content, known as the WEBA instrument (well-being at work), was developed to assess stress, risks, and job skill learning opportunities. This instrument is based on the Karesek (Karasek, 1979) demand–control theory and innovation guidance (Pot et al., 1994).

A broader illustration is the Finnish policy "Work Environment and Well-Being at Work Until 2020," which identifies attainment of well-being through lengthened working life, decreased accidents and occupational diseases, and reduced physical and psychic strain (Policies for the work environment and well-being, 2011).

The Finnish policy of well-being at work aims to encourage workers to have longer work careers:

> This means improving employees' abilities, will, and opportunities to work. Work must be attractive and it must promote employees' health, work ability, and functional capacity. Good and healthy work environments support sustainable development and employees' well-being and improve the productivity of enterprises and the society. (Policies for the work environment and well-being, 2011, p. 5)

This chapter focuses on public policy related to well-being and the application of that policy to workers and the workplace and for the workforce overall. As shown in Figure 11.3, there are work and nonwork threats to, and promoters of, well-being. Preventing workplace hazards and promoting workers' well-being in that context necessitates policies that cross work–nonwork boundaries. This is exemplified in the WHO "healthy workplaces model" (Burton, 2010) and in the National Institute for Occupational Safety and Health Total Worker Health program (Schill and Chosewood, 2013). Although both acknowledge the nonwork influences, they primarily illustrate how to address them in the workplace.

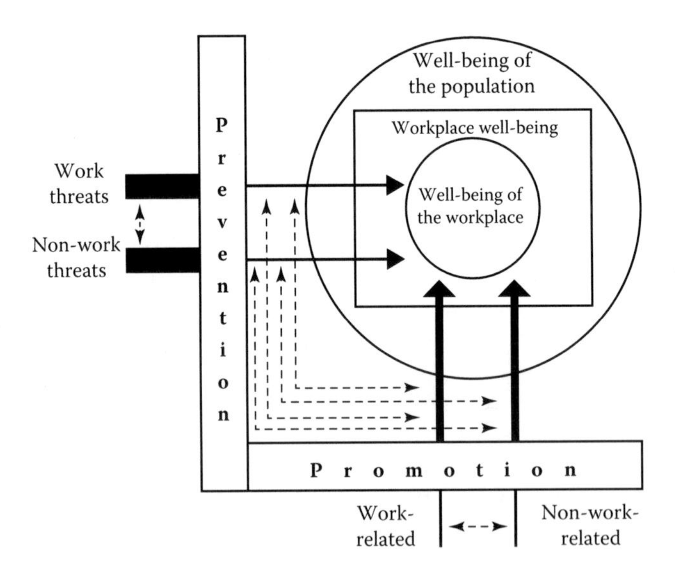

Figure 11.3 Conceptual view of the possible relationship between work and nonwork threats to, and promoters of, well-being. *Note*: Dashed lines show the interaction factors.

Clearly, as has been illustrated in the aftermath of the effort to enact "well-being" legislation in the Netherlands, ensuring well-being in the workplace is difficult to mandate because of various issues related to measurement, responsibility, motivation, and cost (Pot et al., 1994). Nonetheless, there are examples of companies worldwide that have promoted holistic well-being approaches (Dewe and Kompier, 2008; Smith and Clay, 2010; Anttonen, 2009; Delwart et al., 2011; Hammer and Sauter, 2013).

The strong link of well-being to productivity and health is a motivating factor for a business strategy that includes well-being considerations (Rissa and Kaustia, 2007; Harter et al., 2003; Prochaska et al., 2011; Delwart et al., 2011; Hammer and Sauter, 2013; Baptiste, 2008; Loeppke et al., 2009). The key, however, is to integrate health protection and health promotion in interventions in the workplace (Schill and Chosewood, 2013). Although the focus has initially been on physical health, the greater impact may be on mental and social health and well-being overall (Loeppke et al., 2009; Beddington et al., 2008; Black, 2008; Juniper et al., 2009; Harter et al., 2003). Employers have found that it is beneficial to go beyond a physical health focus to affect productivity because recruitment, retention, and engagement of employees include consideration of work–life balance, community involvement, and employee development (Anttonen and Räsänen, 2008; Beddington et al., 2008; Juniper et al., 2009; Harter et al., 2003).

The implementation of programs for well-being at work has been described in various countries and for various sizes of companies (Prochaska et al., 2011; Anttonen and Räsänen, 2008; Beddington et al., 2008; Black, 2008; Juniper et al., 2009). Though generally not explicit about the definition of well-being, the programs involve raising awareness in a company, creating a culture and environment that will promote practice for well-being, and using a means to measure and consider changes in well-being or its components and determinants. One necessary aspect that is not often highlighted in implementing well-being policies is the need for worker participation (Punnett et al., 2013). Worker input on policies and processes that affect them is critical for the effectiveness of those efforts.

From a business perspective, implementation of well-being programs should be considered as an investment rather than a cost (Howard, 2013; Juniper, 2012). However, as Cherniack notes,

> The concept of return on investment for a workplace intervention is incongruous with the more usual approach in health research of comparative effectiveness, where interventions are compared between nonamortized outcomes or comparable outcomes are assessed by program costs. (Cherniack, 2013, p. 44)

Nonetheless, the business value of well-being programs is critical information (Harter et al., 2003; Prochaska et al., 2011; Gandy et al., 2014; Baptiste, 2008; Loeppke et al., 2009). To this end, efforts to develop and implement well-being policies should promote the business value of such policies along with the philosophical and humanitarian aspects (Prochaska et al., 2011; Gandy et al., 2014; Baptiste, 2008; Black, 2008). Well-being at work has implications not only for the worker and the company but also for the workforce as a whole, for the population, and for the national economy (Schulte and Vainio, 2010; Prochaska et al., 2011; Beddington et al., 2008; Black, 2008; Robertson and Cooper, 2011).

Restricting consideration of well-being to the workplace does not capture the full importance and meaning of work in human life. Budd and Spencer have advocated that

> [a] more complete approach to worker well-being needs to go beyond job quality to consider workers as fully-functioning citizens who derive and experience both public and private benefits and costs from working. (Budd and Spencer, 2014, p. 3)

Too often worker well-being is considered at the level of worker health, job quality, and satisfaction. However, interviews with workers about their perceived well-being found that many emphasized specific job characteristics, and many also focused on the extent to which work enabled them to live in their chosen community and attain or maintain their preferred lifestyle (Budd and Spencer, 2014; Cooke et al., 2013). The extent to which this sentiment is generalizable to a broad range of workers is not known, but it is clear that job quality and worker well-being has a socioeconomic context that should be considered in the development of public policy and implementation of well-being programs (Budd and Spencer, 2014; Rath et al., 2010; Blanchflower, 2004).

Research and operationalization needs

Although there is a foundational literature on the relationship between worker well-being and enterprise productivity (Rissa and Kaustia, 2007; Harter et al., 2003; Prochaska et al., 2011; Loeppke et al., 2009; Juniper et al., 2009; Harter et al., 2003; Robertson and Cooper, 2011), the link has yet to be analyzed systematically and bears further investigation (Danna and Griffin, 1999). In addition, the relationship between workforce well-being and population well-being needs to be further elaborated to provide the impetus for the development of policies and practices to support programs that enhance well-being (Schulte and Vainio, 2010). Workforce well-being historically has not always been viewed as central to national welfare (Schulte and Vainio, 2010; Budd and Spencer, 2014). This is despite the fact that the workforce makes up the largest group in the population (when compared with the prework and postwork groups). There is also need for research on the creation of good jobs. A large number of "good" jobs and the ability of the workforce to access them, move between them, and otherwise thrive is critical to the well-being of the entire population (Osterman and Shulman, 2011; Grawitch et al., 2006; Clifton, 2011).

There is also a need for a strategy for conducting research to fill gaps in the evidence base for what works and does not work in achieving or maintaining well-being. These efforts will hinge on clarifying the constituent factors that contribute to well-being, as well as on identifying promising interventions to address or enhance well-being. Many of the challenges related to conceptualizing and operationalizing well-being may prove intractable, including addressing issues linked to distribution of opportunity and income, lack of job control, organization of work, and potential conflicts caused when or if guidance for promoting worker well-being is perceived as interfering with employers' rights to manage their workplaces (Hansen and Starheim, 2011). Nonetheless, as described earlier in this chapter, the Canadian mental health standard shows that many of these barriers are not insurmountable (Canadian Standards Association, Bureau de normalization du Quebec, 2013). An important focus of research is how to address these seemingly intractable challenges linked to promoting well-being. It may be desirable to include well-being in public policy, but several challenges exist to achieving this end. Empirical research and analysis require the development of standard definitions, operationalized variables that can be measured, and an understanding of the implications of using such definitions and metrics (Adler, 2011; Fleuret and Atkinson, 2007; Seligman, 2011; Mueller, 2003; Ilies et al., 2007). Hypothesis testing of the determinants and impacts of well-being, based on the operationalization of definitions and associated metrics, moves the researcher from the abstract to the empirical level, where variables—rather than concepts or definitions—are the focus (Mueller, 2003). If variables that describe aspects of well-being can be measured and agreed upon, this might allow the determination of targets for intervention or improvement. Analysis of the determinants and impacts of well-being for public policy need to be multilevel because well-being is affected by political, economic, and social factors (Ng and Fisher, 2013; Xing and Chu, 2011; Allin, 2014; Siegrist, 1996; Ilies et al., 2007).

One of the drivers of public policy aimed at occupational safety and health is quantitative risk assessment (National Research Council, 2009). Quantitative risk assessment is statistical modeling of hazard exposure–response data to predict risks (usually) at lower levels of exposure. One question is whether approaches from classic quantitative risk assessment are useful in assessing well-being in the occupational setting as a basis for policy. To apply risk assessment methods to well-being, an "exposure–response" analog would be needed. For example, the exposures could be threats to well-being and the response the state of well-being influenced by those threats, or exposures could be a certain type of well-being and the response could be markers of health. A number of issues arise in the conceptualization of an exposure–response relationship related to well-being. First, threats to well-being can come from various sources related to work and external to it. Identifying them and combining them into an exposure variable with an appropriate, tractable metric will be difficult. If the focus is for employer and worker or workforce guidance, it is likely that the threats pertaining to the workplace will be most relevant, but external nonwork threats will also be important, and the two may interact (Figure 11.3). Capturing that interaction will be difficult and complicated but it is critical if well-being is to be included in risk assessment. Second, there are factors that promote well-being, such as work, having a job, and adequate income (Black, 2008; Waddell and Burton, 2006; Stevenson and Wolfers, 2013). A job that is satisfying is even more so associated with well-being (Grawitch et al., 2006; Broom et al., 2006). Third, identifying and operationalizing the promoters of or threats to well-being will be a challenge.

If this can be accomplished, then perhaps exposure–response relationships relevant to well-being can be identified and quantified. It may then be appropriate to identify on an "exposure–response curve" an (adverse) change in the "amount" of well-being, or other outcome affected by well-being, that may be considered to be of importance in a given context. This may then allow the derivation of a level of a relevant exposure associated with the defined change in outcome, consistent with the application of the quantitative risk assessment paradigm. If exposures (positive and negative) could be assessed, then desirable targets could be specified. Ultimately, the need will be to determine a level of risk for either decreased well-being or outcomes impacted by well-being in the workplace, the workforce, and for individual workers to define targets for intervention and prevention strategies. If a risk assessment approach is appropriate to examine well-being for the development of public policy, one underlying question is whether it is feasible to develop an exposure–response analog for well-being. Various aspects of well-being would need to be taken into consideration. These include the fact that well-being has both subjective and objective attributes. How do you measure these across various work settings and conditions? (Spector et al., 2002; Trimble, 2010) How do you adjust for subjective differences? (Diener, 2013) In addition, well-being is not a static condition; it is an evolving one that changes with time and other factors (Fleuret et al., 2007; Sointu, 2005). Although the focus thus far has been on well-being as the target of policy, it may be that well-being also should be seen as a means to achieve policy (Atkinson, 2012). If the latter view is the case, does this change any of the definitional and variable specification issues raised thus far? For example, the issue of well-being as a changing condition might be the rationale for identifying trends in workforce well-being as outcomes of policies or interventions. As a consequence, leading indicators of well-being outcomes might be sought and used as targets for guidance or regulation. If well-being is to be incorporated in quantitative risk assessments, it would need to be evaluated for whether it meets the basic quantitative risk assessment criteria. These include whether the hazards (threats) to well-being can be defined and measured; the well-being can be defined and measured when it is functioning as a factor that has an impact on health outcomes; the well-being as an outcome or response of exposure to these hazards (threats) can be defined and measured; the exposure–response models are appropriate for measuring well-being and whether these need to be quantitative, qualitative, or a combination thereof; the risk can be characterized and if it is possible to account for aspects such as uncertainty and sensitive populations; and the risk characteristics of threats to well-being can be used to drive risk-management strategies. It may be that the exposure–response paradigm is not the appropriate way to think about and assess well-being. Qualitative approaches may be more relevant. For example, the European Union Well-Being at Work project promoted an approach developed by the Finnish Institute of Occupational Health, which described the management changes needed to achieve well-being at work and provided a self-evaluation matrix with six categories of activities that businesses can use to evaluate their performance (Anttonen and Räsänen, 2008; Anttonen and Vainio, 2010). Or perhaps a combined quantitative and qualitative approach is the most appropriate for determining, addressing, or measuring the well-being of worker populations.

Going forward

The ultimate policy issue is whether it is a good idea to incorporate well-being as a focus for occupational risk assessments and guidance. The benefits of operationalizing well-being need further investigation, as do the unintended consequences. There are various policy issues that might come into play. Some of these have been described, including "blaming the worker" for lack of well-being in the workplace and diluting the responsibility of employers. The means of achieving well-being in the workplace are generally the responsibility of the employer, but true well-being requires that workers have autonomy and a role in determining the organization and conditions of work, have a living wage, and have an opportunity to share in the success of the organization. Clearly, these will be viewed by some as controversial issues, but they stem from what has been averred as inherent rights, even if they are not universally recognized (Osterman and Shulman, 2011; Bauer, 2010; Budd and Spencer, 2014; Hodson, 2001; Fisk, 2007; UN General Assembly, 1948).

Eventually, consideration of well-being has to address ethical issues because various definitions of well-being could involve questions such as fair distribution of opportunity and realization of self-determination. It is understandable that there are political differences that arise over these issues. To help bridge some of these differences, there is need to expand the knowledge base described in this chapter for such aspects as the business case for promoting well-being at work, the link between workforce well-being and population well-being, and the need to address the range of determinants of well-being. The challenge in making workforce well-being a focus of public health and ultimately societal expectation is that it requires multiple disciplines and stakeholder groups to interact, communicate, and ultimately work together. This is not easy to achieve. In fact, it is quite difficult.

There is need for development of a strategy for cross-discipline and cross-stakeholder group multiway communication on this issue. The specialty discipline of occupational safety and health may be an appropriate initiator of these communications because its focus is solidly rooted in workforce safety, health, and, by extension, well-being. Other disciplines such as occupational health psychology, health promotion, industrial and organizational psychology, ergonomics, economics, sociology, geography, medicine, law, nursing, and public health also should be involved in the effort to examine well-being guidance and policies. The focus on worker and workforce well-being is of critical national importance because of the role and significance of work to national life (Schulte and Vainio, 2010; Loeppke et al., 2009; Black, 2008; Anttonen and Vainio, 2010). The growing incidence of mental disorders, stress-related outcomes, and chronic diseases in the population and the organizational features of work related to safety and health outcomes require attention to their linkage to well-being.

With dependency ratios becoming dangerously close to unsustainable levels and rising health care burdens, there is need for a more comprehensive view of what factors deleteriously affect the workforce and what can be done about it. Considering and operationalizing the concept of well-being is an important next step.

About the authors

Paul A. Schulte, Rebecca J. Guerin, Anasua Bhattacharya, Thomas R. Cunningham, Sudha P. Pandalai, Donald Eggerth, and Carol M. Stephenson are with the Education and Information Division, the National Institute for Occupational Safety and Health, Cincinnati, OH. Anita L. Schill is with the Office of the Director, National Institute for Occupational Safety and Health, Washington, DC.

Correspondence should be sent to Paul A. Schulte, PhD, the National Institute for Occupational Safety and Health, 1150 Tusculum Ave, MS-C14, Cincinnati, OH 45226 (e-mail: pschulte@cdc.gov). Reprints can be ordered at http://www.ajph.org by clicking the "Reprints" link. This article was accepted January 30, 2015.

Note: The findings and conclusions in this report are those of the author(s) and do not necessarily represent the views of the National Institute for Occupational Safety and Health.

Contributors

P. A. Schulte originated the document, developed the first draft, and revised subsequent drafts. R. J. Guerin reviewed definitions of well-being, contributed to writing the article, and edited various drafts. A. L. Schill, A. Bhattacharya, S. P. Pandalai, T. R. Cunningham, D. Eggerth, and C.M. Stephenson contributed to writing and editing the article.

Acknowledgments

We thank the following individuals for comments on earlier versions of this article: Frank Pot, Knut Ringen, Gregory Wagner, and Laura Punnett. We also acknowledge the efforts of Brenda Proffitt, Elizabeth Zofkie, and Devin Baker regarding logistical support in preparation of this article.

Human participant protection

No institutional review board approval was needed because the focus was literature based.

References

Adler M. *Well-Being and Equity: A Framework for Policy Analysis*. New York: Oxford University Press; 2011.

Adler MD. Happiness surveys and public policy: What's the use? *Duke Law J*. 2013;62(8):1509–1601.

Agrawal S, Harter JK. *Well-Being Meta-analysis: A Worldwide Study of the Relationship between the Five Elements of Wellbeing and Life Evaluation, Daily Expenses, Health, and Giving*. Washington, DC: Gallup; 2001.

Allin P. Measuring well-being in modern societies. In: Chen PY, Cooper CL, eds. *Work and Wellbeing: A Complete Reference Guide*. West Sussex, UK: John Wiley; 2014.

Antonovsky A. The salutogenic model as a theory to guide health promotion. *Health Promot Int*. 1996;11(1):11–18.

Anttonen H, Räsänen T. *EU and Well-Being at Work Policy: Well-Being at Work in Partner Countries*. Helsinki, Finland: Finnish Institute of Occupational Health; 2008.

Anttonen H, Vainio H. Towards better work and well-being an overview. *J Occup Environ Med*. 2010;52 (12):1245–1248.

Anttonen HT. *Well-Being at Work: New Innovations and Good Practices*. Helsinki, Finland: Finnish Institute of Occupational Health; 2009:32.

Atkinson S, Fuller S, Painter J. *Well-Being and Place*. Surrey, UK: Ashgate; 2012.

Atkinson S, Joyce Kerry E. The place and practices of wellbeing in local governance. *Environ Plann C Gov Policy*. 2011;29:133–168.

Bakker AB, Demerouti E. The job demands–resources model: State of the art. *J Manag Psychol*. 2007;22(3):309–328.

Baptiste NR. Tightening the link between employee wellbeing at work and performance: A new dimension for HRM. *Manage Decis*. 2008;46(1–2):284–309.

Bauer GF. Determinants of workplace health: Salutogenic perspective on work, organization, and organizational change. Slide presentation at: Exploring the SOC determinants for health, 3rd International Research Seminar on Salutogenesis and the 3rd Meeting of the International Union for Health Promotion and Education Global Working Group on Salutogenesis; July 11, 2010; Geneva, Switzerland. Available at: http://www. salutogenesis.hv.se/files/bauer_g._genf.salutogen.wh.pdf. Accessed March 6, 2015.

Beddington J, Cooper CL, Field J, et al. The mental wealth of nations. *Nature*. 2008;455(7216):1057–1060.

Belgian Legislation. Act of August 4, 1996 on well-being of workers in the performance of their work. Available at: http://www.beswic.be/fr/legislation/the-belgian-legislation/front-page. Accessed March 6, 2015.

Black C. *Working for a Healthier Tomorrow*. London: The Stationary Office; 2008.

Blanchflower DG, Oswald AJ. Well-being over time in Britain and the USA. *J Public Econ*. 2004;88(7–8):1359–1386.

Blustein D. *The Psychology of Working*. Mahwah, NJ: Lawrence Erlbaum; 2006.

Bockerman P, Ilmakunnas P, Johansson E. Job security and employee well-being: Evidence from matched survey and register data. *Labour Econ*. 2011; 18(4):547–554.

Boehm JK, Peterson C, Kivimaki M, Kubzansky L. A prospective study of positive psychological well-being and coronary heart disease. *Health Psychol*. 2011;30(3):259–267.

Bronsteen J, Buccafusco CJ, Masur JS. *Well-Being and Public Policy*. Chicago, IL: University of Chicago Coase–Sandor Institute for Law and Economics; 2014. Research paper 707.

Broom DH, D'Souza RM, Strazdins L, Butterworth P, Parslow R, Rodgers B. The lesser evil: Bad jobs or unemployment? A survey of mid-aged Australians. *Soc Sci Med*. 2006;63(3):575–586.

Budd JW, Spencer DA. Worker well-being and the importance of work: Bridging the gap. *Euro J Indust Rel*. 2014:1–16.

Buffet MA, Gervais RL, Liddle M, Eechelaert L. *Well-Being at Work: Creating a Positive Work Environment*. Luxembourg: European Agency for Safety and Health at Work; 2013.

Burton J. *Healthy Workplaces: A Model for Action: For Employers, Workers, Policy-Makers, and Practitioners*. Geneva, Switzerland: World Health Organization; 2010.

Bustillos AS, Ortiz Trigoso O. Access to health programs at the workplace and the reduction of work presenteeism: A population-based cross-sectional study. *J Occup Environ Med*. 2013;55(11):1318–1322.

Canadian Standards Association, Bureau de normalization du Quebec. *Psychological Health and Safety in the Workplace: Prevention, Promotion, and Guidance to Staged Implementation*. Toronto, ON: Mental Health Commission of Canada, National Standard of Canada; 2013.
Cherniack M. Integrated health programs, health outcomes, and return on investment: Measuring workplace health promotion and integrated program effectiveness. *J Occup Environ Med*. 2013;55(12 suppl):S38–S45.
Chida Y, Steptoe A. Positive psychological well-being and mortality: A quantitative review of prospective observational studies. *Psychosom Med*. 2008;70(7):741–756.
Church TS, Thomas DM, Tudor-Locke C, et al. Trends over five decades in US occupation-related physical activity and their associations with obesity. *PLoS ONE*. 2011;6(5):e19657.
Cierpich H, Styles L, Harrison R, et al. Work-related injury deaths among Hispanics: United States, 1992–2006 (Reprinted from MMWR, Vol. 57, pp. 597–600, 2008). *JAMA*. 2008;300(21):2479–2480.
Clark AE. Job satisfaction in Britain. *Br J Ind Relat*. 1996;34(2):189–217.
Clifton J. *The Coming Jobs War*. New York: Gallup Press; 2011.
Cohen GA. Equality of what? On welfare, goods and capabilities. In: Nussbaum M, Sen AK, eds. *The Quality of Life*. Oxford, UK: Clarendon Press; 1993.
Cooke GB, Donaghey J, Zeytinoglu IU. The nuanced nature of work quality: Evidence from rural Newfoundland and Ireland. *Hum Relat*. 2013;66(4):503–527.
Costa G, Goedhard WJ, Ilmarinen J. *Assessment and Promotion of Work Ability, Health and Well-Being of Ageing Workers*. Amsterdam, the Netherlands: Elsevier; 2005.
Cronin de Chavez A, Backett-Milburn K, Parry O, Platt S. Understanding and researching wellbeing: Its usage in different disciplines and potential for health research and health promotion. *Health Educ J*. 2005;64(1):70–87.
Cummings KJ, Kreiss K. Contingent workers and contingent health: Risks of a modern economy. *JAMA*. 2008;299(4):448–450.
Danna K, Griffin RW. Health and well-being in the workplace: A review and synthesis of the literature. *J Manage*. 1999;25(3):357–384.
Dawis RV, Lofquist LH. *A Psychological Theory of Work Adjustment: An Individual-Differences Model and Its Applications*. Minneapolis: University of Minnesota Press; 1984.
de Perio MA, Wiegand DM, Brueck SE. Influenza-like illness and presenteeism among school employees. *Am J Infect Control*. 2014;42(4):450–452.
Deaton A, Stone AA. Two happiness puzzles. *Am Econ Rev*. 2013;103(3):591–597.
Delwart V, Van Eck T, Vandehende A, Berruga B. *Wellbeing in the Workplace: A Guide with Best Practices* and *Tips for Implementing a Successful Wellbeing Strategy at Work*. Brussels, Belgium: CSR Europe; 2011. Available at: http://www.csreurope.org/sites/default/files/wellbeingguide2_0.pdf. Accessed March 6, 2015.
Deming WE. *Out of the Crisis 2000*. Cambridge, MA: Center for Advanced Manufacturing Study; 2008.
DeVol R, Bedroussian A, Charuworn A, et al. *An Unhealthy America: The Economic Burden of Chronic Disease*. Santa Monica, CA: Milken Institute; 2007.
Dewe P, Kompier M. Foresight mental capital and wellbeing project. In: Dewe P, Kompier M, eds. *Wellbeing and Work: Future Challenges*. London: Government Office for Science; 2008.

Dhondt S, Pot F, Kraan K. The importance of organizational level decision latitude for well-being and organizational commitment. *Team Perform Manage*. 2014;20:307–327.

Diaz-Serrano L, Cabral Vieira J. *Low-Pay, Higher Pay and Job Satisfaction within the European Union Empirical Evidence from Fourteen Countries*. Bonn, Germany: Institute for the Study of Labor; 2005. IZA Discussion Papers no. 1550.

Diener E. The remarkable changes in the science of subjective well-being. *Perspect Psychol Sci*. 2013;8(6):663–666.

Diener E, Chan MY. Happy people live longer: Subjective well-being contributes to health and longevity. *Appl Psychol*. 2011;3(1):1–43.

Diener E, Oishi S, Lucas RE. Personality, culture, and subjective well-being: Emotional and cognitive evaluations of life. *Annu Rev Psychol*. 2003;54:403–425.

Dolan P, White MP. How can measures of subjective well-being be used to inform public policy? *Perspect Psychol Sci*. 2007;2(1):71–85.

East Riding of Yorkshire Council. Well-being at work code of practice. 2004. Available at: http://www.eriding.net/resources/educators/well-being/070316_rmciver_edu_ssu27_code_of_practice.pdf. Accessed March 6, 2015.

Easterlin RA. Does economic growth improve the human lot? In: David PA, Reeder MW, eds. *Nations and Households in Economic Growth: Essays in Honor of Moses Abramovitz*. New York: Academic Press; 1974:89–125.

ESF Age Network. Work Ability Index: A programme from the Netherlands. Available at: http://www.careerandage.eu/prevsite/sites/esfage/files/resources/Work-Ability-Index_1.pdf. Accessed March 4, 2015.

European Working Conditions Observatory. *Well-Being at Work: Innovation and Good Practice*. Dublin, Ireland: European Foundation for the Improvement of Living and Working Conditions; 2008.

Fisk M. Human rights and living wages. In: Butler C, ed. *Guantanamo Bay and the Judicial-Moral Treatment of the Other*. West Lafayette, IN: Purdue University Press; 2007:146–160.

Fleuret S, Atkinson S. Wellbeing, health and geography: A critical review and research agenda. *NZ Geog*. 2007;63(2):106–118.

Flynn MA. Safety and the diverse workforce. *Prof Saf*. 2014;59:52–57.

Frey BS, Gallus J. Subjective well-being and policy. *Topoi*. 2013;32(2):207–212.

Frey BS, Stutzer A. *Happiness and Economics: How the Economy and Institutions Affect Human Well-Being*. Princeton, NJ: Princeton University Press; 2002.

Gandy WM, Coberley C, Pope JE,Wells A, Rula EY. Comparing the contributions of well-being and disease status to employee productivity. *J Occup Environ Med*. 2014;56(3):252–257.

Gaspart F. Objective measures of well-being and the cooperative production problem. *Soc Choice Welfare*. 1997;15(1):95–112.

Grawitch MJ, Gottschalk M, Munz DC. The path to a healthy workplace: A critical review linking healthy workplace practices, employee well-being, and organizational improvements. *Consult Psychol J Pract Res*. 2006; 58(3):129–147.

Grebner S, Semmer NK, Elfering A. Working conditions and three types of well-being: A longitudinal study with self-report and rating data. *J Occup Health Psychol*. 2005;10(1):31.

Guillemin M. *Lesser Known Aspects of Occupational Health* [in French]. Paris, France: L'Harmattan; 2011.

Hammer LB, Sauter S. Total worker health and work-life stress. *J Occup Environ Med*. 2013;55(12 suppl):S25–S29.
Hansen AM, Starheim L. Working environment authority interventions to promote well-being at work. Paper presented at: 29th International Labour Process Conference; April 5–7, 2011; Leeds, UK.
Harter JK, Schmidt FL, Keyes CL. Well-being in the workplace and its relationship to business outcomes: A review of the Gallup studies. In: Keyes CL, Haidt J, eds. *Flourishing: Positive Psychology and the Life Well-Lived*. Washington, DC: American Psychology Association; 2003:205–224.
Harter JK, Schmidt FL, Killham EA. *Employee Engagement, Satisfaction, and Business-Unit-Level Outcomes: A Meta-analysis*. Princeton, NJ: Gallup Organization; 2003.
Hassan E, Austin C, Celia C, et al. *Health and Well-Being at Work in the United Kingdom*. Cambridge, UK: Rand Europe; 2009.
Hemp P. Presenteeism: At work—but out of it. *Harv Bus Rev*. 2004;82(10):49–58.
Hendrick HW, Kleiner B. *Macroergonomics: Theory, Methods, and Applications*. Mahwah, NJ: CRC Press; 2005.
Hodson R. *Analyzing Documentary Accounts* (Quantitative Applications in the Social Sciences). Thousand Oaks, CA: Sage; 1999.
Hodson R. *Dignity at Work*. Cambridge, UK: Cambridge University Press; 2001.
Holzer HJ, Nightingale DS. *Reshaping the American Workforce in a Changing Economy*. Washington, DC: The Urban Institute; 2007.
Howard J. Seven challenges for the future of occupational safety and health. *J Occup Environ Hyg*. 2010;7(4):D11–D18.
Howard J. Worker health = economic health. In: *27th National Conference on Health, Productivity, and Human Capital*. Washington, DC: National Business Group on Health; 2013.
Howell RT, Kern ML, Lyubomirsky S. Health benefits: Meta-analytically determining the impact of well-being on objective health outcomes. *Health Psychol Rev*. 2007;1(1):83–136.
Huber M, Knottnerus JA, Green L, et al. How should we define health? *BMJ*. 2011;343:d4163.
Huppert FA. Psychological well-being: Evidence regarding its causes and consequences. *Appl Psychol*. 2009;1(2):137–164.
Huppert FA, Marks N, Clark A, et al. Measuring well-being across Europe: Description of the ESS wellbeing module and preliminary findings. *Soc Indic Res*. 2009;91(3):301–315.
Ilies R, Schwind KM, Heller D. Employee well-being: A multilevel model linking work and nonwork domains. *Eur J Work Organ Psychol*. 2007;16(3):326–341.
Ilmarinen J, Tuomi K, Eskelinen L, Nygård C-H, Huuhtanen P, Klockars M. Summary and recommendations of a project involving cross-sectional and follow-up studies on the aging worker in Finnish municipal occupations (1981–1985). *Scand J Work Environ Health*. 1991;16(suppl 1):135–141.
Johns G. Presenteeism in the workplace: A review and research agenda. *J Organ Behav*. 2010;31(4):519–542.
Johnson JV, Hall EM. Job strain, work place social support, and cardiovascular-disease: A cross-sectional study of a random sample of the Swedish working population. *Am J Public Health*. 1988;78(10):1336–1342.
Juniper B. Paying the price for well-being. *Occup Health* (Lond). 2012;2012:25–26.

Juniper B, White N, Bellamy P. Assessing employee wellbeing: Is there another way? *Int J Workplace Health Manag*. 2009;2(3):220–230.

Kahneman D. Objective happiness. In: Kahneman D, Diener E, Schwarz N, eds. *Well-Being: Foundations of Hedonic Psychology*. New York: Russell Sage Foundation; 2003:3–25.

Karanika-Murray M, Weyman AK. Optimising workplace interventions for health and well-being: A commentary on the limitations of the public health perspective within the workplace health arena. *Int J Workplace Health Manag*. 2013;6(2):104–117.

Karasek R. *Healthy Work: Stress, Productivity, and the Reconstruction of Working Life*. New York: Basic Books; 1992.

Karasek RA. Job demands, job decision latitude, and mental strain: Implications for job redesign. *Adm Sci Q*. 1979;24(2):285–308.

Kitt M, Howard J. The face of occupational safety and health: 2020 and beyond. *Public Health Rep*. 2013;128 (3):138–139.

Kochan T, Finegold D, Osterman P. Who can fix the "middle-skills" gap? *Harv Bus Rev*. 2012;90:83–90.

Kompier MA. New systems of work organization and workers' health. *Scand J Work Environ Health*. 2006;32(6):421–430.

Linley PA, Maltby J, Wood AM, Osborne G, Hurling R. Measuring happiness: The higher order factor structure of subjective and psychological well-being measures. *Pers Individ Dif*. 2009;47(8):878–884.

Loeppke R, Taitel M, Haufle V, Parry T, Kessler RC, Jinnett K. Health and productivity as a business strategy: A multiemployer study. *J Occup Environ Med*. 2009;51(4):411–428.

Mellor N, Webster J. Enablers and challenges in implementing a comprehensive workplace health and well-being approach. *Int J Workplace Health Manag*. 2013;6:129–142.

Michaelson J, Abdallah S, Steuer N, et al. *National Accounts of Well-Being: Bringing Real Wealth onto the Balance Sheet*. London: New Economics Foundation; 2009.

Mueller CW. Conceptualization, operationalization, and measurement. In: Lewis-Beck MS, Bryman A, Liao TF, eds. *Encyclopedia of Social Science Research Methods*. Thousand Oaks, CA: Sage; 2003.

National Institute for Occupational Safety and Health. Essential elements of effective workplace programs and policies for improving worker health and wellbeing. Publication no. 2010–140, Washington, DC: NIOSH; 2010.

National Research Council. *Science and Decisions: Advancing Risk Assessment*. Washington, DC: The National Academies Press; 2009.

National Research Council. Subjective well-being: Measuring happiness, suffering, and other dimensions of experience. In: Stone AA, Machie E, eds. *Paper on Measuring Subjective Well-Being in a Policy Framework*. Washington, DC: National Academies Press; 2013.

Neubauer D, Pratt R. The 2nd public-health revolution: A critical-appraisal. *J Health Polit Policy Law*. 1981;6(2):205–228.

Ng ECW, Fisher AT. Understanding well-being in multi-levels: A review. *Health Cult Sc*. 2013;5(1):308–323.

Noor NM. Work- and family-related variables, work–family conflict and women's well-being; Some observations. *Community Work Fam*. 2003;6(3):297–319.

Osberg L. The relevance of objective indicators of well-being for public policy. In: *CSLS Session on Well-Being at the Annual Meeting of the Canadian Economics Association*. Toronto, Ontario: Ryerson University; 2004:2–4.

Osterman P, Shulman B. *Good Jobs America*. New York: Russell Sage Foundation; 2011.

Otoiu A, Titan E, Dumitrescu R. Are the variables used in building composite indicators of well-being relevant? Validating composite indexes of well-being. *Ecol Indic*. 2014;46:575–585.

Pandalai SP, Schulte PA, Miller DB. Conceptual heuristic models of the interrelationships between obesity and the occupational environment. *Scand J Work Environ Health*. 2013;39(3):221–232.

Passel JS, Cohn DV. *US Population Projections: 2005–2050*. Washington, DC: Pew Research Center; 2008.

Pot FD, Peeters MHH, Vaas F, Dhondt S. Assessment of stress risks and learning opportunities in the work organization. *Euro Work and Organ Psychol*. 1994;4(1):21–37.

Pressman SD, Cohen S. Does positive affect influence health? *Psychol Bull*. 2005;131(6):925–971.

Prochaska JO, Evers KE, Johnson JL, et al. The well-being assessment for productivity: A well-being approach to presenteeism. *J Occup Environ Med*. 2011; 53(7):735–742.

Punnett L, Warren N, Henning R, Nobrega S, Cherniack M, CPH-NEW Research Team. Participatory ergonomics as a model for integrated programs to prevent chronic disease. *J Occup Environ Med*. 2013;55(12 suppl):S19–S24.

Rajaratnam AS, Sears LE, Shi Y, Coberley CR, Pope JE. Well-being, health, and productivity improvement after an employee well-being intervention in large retail distribution centers. *J Occup Environ Med*. 2014;56 (12):1291–1296.

Rath T, Harter J, Harter JK. *Wellbeing: The Five Essential Elements*. New York: Gallup Press; 2010.

Rawls J. *A Theory of Justice*. Cambridge, MA: Harvard University Press; 2009.

Ringen S. Households, goods, and well-being. *Rev Income Wealth*. 1996;4:421–431.

Rissa K, Kaustia T. *Well-Being Creates Productivity: The Druvan Model*. Iisalmi, Finland: Centre for Occupational Safety and the Finnish Work Environment Fund; 2007.

Robertson I, Cooper CL. *Well-Being: Productivity and Happiness at Work*. New York: Palgrave Macmillan; 2011.

Rounds JB, Dawis RV, Lofquist LH. Measurement of person environment fit and prediction of satisfaction in the theory of work adjustment. *J Vocat Behav*. 1987;31 (3):297–318.

Ryan RM, Deci EL. On happiness and human potentials: A review of research on hedonic and eudaimonic well-being. *Annu Rev Psychol*. 2001;52:141–166.

Ryff CD. Psychological well-being revisited: Advances in the science and practice of eudaimonia. *Psychother Psychosom*. 2014;83(1):10–28.

Schill AL, Chosewood LC. The NIOSH total worker health program: An overview. *J Occup Environ Med*. 2013;55(12 suppl):S8–S11.

Schulte P, Vainio H. Well-being at work: Overview and perspective. *Scand J Work Environ Health*. 2010; 36(5):422–429.

Schulte PA. Emerging issues in occupational safety and health. *Int J Occup Environ Health*. 2006;12(3):273–277.

Schulte PA, Pandalai S, Wulsin V, Chun H. Interaction of occupational and personal risk factors in workforce health and safety. *Am J Public Health*. 2012;102 (3):434–448.

Seligman MEP. *Flourish: A Visionary New Understanding of Happiness and Well-Being*. New York: Simon and Schuster; 2011.

Sen A. Capability and well-being. In: Nussbaum M, Sen AK, eds. *The Quality of Life*. Oxford, UK: Clarendon Press; 1993:30–54.

Sen A. *Development as Freedom*. New York: Oxford University Press; 1999.

Shrestha LB. *Age Dependency Ratios and Social Security Solvency*. Washington, DC: CRS Report for Congress; 2006.

Siegrist J. Adverse health effects of high-effort/low-reward conditions. *J Occup Health Psychol*. 1996;1(1):27–41.

Smith CL, Clay PM. Measuring subjective and objective well-being: Analyses from five marine commercial fisheries. *Hum Organ*. 2010;69(2):158–168.

Sointu E. The rise of an ideal: Tracing changing discourses of wellbeing. *Sociol Rev*. 2005;53(2):255–274.

Sparks K, Faragher B, Cooper CL. Well-being and occupational health in the 21st century workplace. *J Occup Organ Psychol*. 2001;74(4):489–509.

Spector PE, Cooper CL, Sanchez JI, et al. Locus of control and well-being at work: How generalizable are Western findings? *Acad Manage J*. 2002;45(2):453–466.

Steptoe A, Wardle J, Marmot M. Positive affect and health-related neuroendocrine, cardiovascular, and inflammatory processes. *Proc Natl Acad Sci USA*. 2005;102(18):6508–6512.

Stevenson B, Wolfers J. *Subjective Well-Being and Income: Is There Any Evidence of Satiation?* Cambridge, MA: National Bureau of Economic Research; 2013.

Stiglitz JE, Sen A, Fitoussi J-P. Report by the Commission on the Measurement of Economic Performance and Social Progress. 2009. Available at: http://www.stiglitz-sen-fitoussi.fr/documents/rapport_anglais.pdf. Accessed March 6, 2015.

Stone KV. *From Widgets to Digits: Employment Regulation for the Changing Workplace*. New York: Cambridge University Press; 2004.

Tehrani N, Humpage S, Willmott B, Haslam I. *What's Happening with Well-Being at Work?* London: Chartered Institute of Personnel and Development; 2007.

Trimble JE. Cultural measurement equivalence. In: Clauss-Ehlers C, ed. *Encyclopedia of Cross-Cultural School Psychology*. New York: Springer; 2010:316–318.

UN General Assembly. Universal Declaration of Human Rights, December 10, 1948: 217 A (III). Available at: http://www.refworld.org/docid/3ae6b3712c.html. Accessed January 26, 2015.

Vainio H. Salutogenesis to complement pathogenesis: A necessary paradigm shift for well-being at work. Slide presentation at: 30th International Congress on Occupational Health; March 18–23, 2012; Cancun, Mexico.

Waddell G, Burton AK. *Is Work Good for Your Health and Well-Being?* London: The Stationery Office; 2006.

Warr P. *Work, Unemployment, and Mental Health*. New York: Oxford University Press; 1987.

Warr P. The measurement of well-being and other aspects of mental-health. *J Occup Psychol*. 1990;63(3):193–210.

Warr P. Well-being and the workplace. In: Kahneman D, Diener E, Schwarz N, eds. *Well-Being: Foundations of Hedonic Psychology*. New York: Russell Sage Foundation; 1999:392–412.

Warr P. How to think about and measure psychological well-being. In: Sinclair RR, Wang M, Tetrick L, eds. *Research Methods in Occupational Health Psychology: Measurement Design and Data Analysis*. New York: Routledge Academics; 2012:76–90.

Waterstone ME. Returning veterans and disability law. *Notre Dame Law Rev.* 2010;85:1081–1132.

WHO. WHO definition of health. Geneva, Switzerland: World Health Organization; 1948. Available at: http://www.who.int/about/definition/en/print.html. Accessed March 6, 2015.

Xing Z, Chu L. Research on constructing composite indicator of objective well-being from China mainland. In: *Proceedings of the 58th World Statistical Congress*. Dublin, Ireland: International Statistical Institute; 2011:3081–3097.

Appendix A: Glossary of work performance terms

Acceptance The decision that an item, process, or service conforms to specified characteristics defined in codes, standards, or other requirement documents.

Accountability Responsibility for an activity, accompanied by rewards and recognition for good performance and adverse consequences for performance that is unreasonably poor.

Activity Actions taken by a program or an organization to achieve its objectives.

Assessment An all-inclusive term used to denote the act of determining, through a review of objective evidence and witnessing the performance of activities, whether items, processes, or services meet specified requirements. Assessments are conducted through the implementation of the following actions: audits, performance evaluations, management system reviews, peer reviews, or surveillances, which are planned and documented by trained and qualified personnel.

Assessment/verification The act of reviewing, inspecting, testing, checking, conducting surveillances, auditing, or otherwise determining and documenting whether items, processes, or services meet specified requirements. The US Department of Energy (DOE) order 5700.6C uses the terms *assessment* and *verification* synonymously. This order defines these terms by who is performing the work; assessments are performed by or for senior management, and verifications are performed by the line organization.

Audit A planned and documented activity performed to determine by investigation, examination, or evaluation of objective evidence the adequacy of and compliance with established procedures, instructions, drawings, and other applicable documents and the effectiveness of implementation. Audits should not be confused with surveillances or inspection activities performed for the sole purpose of process control or product acceptance.

Also, a systematic check to determine the quality of operation of some function or activity. Audits may be of two basic types: (1) performance audits in which quantitative data are independently obtained for comparison with routinely obtained data in a measurement system, or (2) system audits of a qualitative nature that consist of an on-site review of a laboratory's quality system and physical facilities for sampling, calibration, and measurement.

Baseline The current level of performance at which an organization, process, or function is operating.

Benchmarking To measure an organization's products or services against the best existing products or services of the same type; the benchmark defines the 100% mark on the measurement scale.

Also, the process of comparing and measuring an organization's own performance in a particular process with the performance of organizations judged to be the best of a comparable industry.

Bottom up Starting with input from the people who actually do the work and consolidating that input through successively higher levels of management.

C-chart Also referred to as a *count chart*, this is used in dealing with counts of a given event over consecutive periods of time. Many of the initial DOE performance indicators involve counts of events for consecutive calendar year quarters, making C-chart analysis of these indicators appropriate.

Cascaded down Starting with a top-level management, communicated to successively lower levels of management and employees.

Characteristics Any property or attribute of an item, process, or service that is distinct, describable, and measurable.

Common causes of variation Indicated by statistical techniques, but the causes themselves need more detailed analysis to be fully identified. Common causes of variation are usually the responsibility of management to correct, although other people directly connected with the process are sometimes in a better position to identify the causes and pass them on to management for correction.

Continuous improvement The undying betterment of a process based on the constant measurement and analysis of results produced by the process and the use of that analysis to modify the process.

Also, where performance gains are maintained and the early identification of deteriorating environmental, safety, and health conditions is accomplished.

Control charts The two main uses for these charts are to monitor whether the system is stable and under control (to warn of changes) and to substantiate results from changes introduced into the system (to confirm positive results).

Control lines/limits The *limit lines* drawn on charts to provide guides for the evaluation of performance indicate the dispersion of data on a statistical basis and indicate whether an abnormal situation (e.g., the process is not in control or special causes are adversely influencing a process in control) has occurred.

Also, two control limits are the statistical mean (average) plus three times the standard deviation and the statistical mean minus three times the standard deviation.

Corrective action Measures taken to rectify conditions adverse to quality and, where necessary, to preclude repetition.

Criteria The rules or tests against which the quality of performance can be measured. They are most effective when expressed quantitatively. Fundamental criteria are contained in policies and objectives, as well as codes, standards, regulations, and recognized professional practices.

Data Factual information, regardless of media and format, used as a basis for reasoning, discussion, or calculation.

Data reduction Any and all processes that change either the form of expression, the quantity of data values, or the numbers of data items.

Data validation The systematic effort to review data in order to ensure acceptable data quality. A systematic process for reviewing a body of data against a set of criteria to provide assurance that the data are adequate for their intended use. A systematic review process conducted to confirm the degree of truth in an analytical measurement.

Distribution charts Data are divided into categories of interest (e.g., root causes or reporting elements). It is then graphed as a stacked bar chart to compare the relative contribution of each category with the total.

Distribution diagram A block diagram showing data in order of their contribution to the total. The horizontal axis of the distribution diagram lists the most frequent item in the performance indicator population on the left and progresses in descending order to the least frequent item on the extreme right. The cumulative total of the items is reflected above the block at each interval. By structuring the data in this form, the distribution diagram highlights the largest contributing items in each performance indicator.

Goal The result that a program or organization aims to accomplish.

Also, a proposed statement of attainment/achievement, with the implication of sustained effort and energy.

Guideline A suggested practice that is not mandatory in programs intended to comply with a standard. The words "should" or "may" denote a guideline; the words "shall" or "must" denote a requirement.

Item An all-inclusive term used in place of the following: appurtenance, sample, assembly, component, equipment, material, module, part, structure, subassembly, subsystem, unit, documented concepts, or data.

Lessons learned A "good work practice" or innovative approach that is captured and shared to promote repeat application. A lesson learned may also be an adverse work practice or experience that is captured and shared to avoid recurrence.

Limit lines Lines drawn on charts to provide guides for the evaluation of performance.

Line manager Includes all managers in the chain of command from the first-line supervisors to the top manager.

Management All individuals directly responsible and accountable for planning, implementing, and assessing work activities.

Mean The arithmetic average of a set of numbers.

Measurement The quantitative parameter used to ascertain the degree of performance.

Metric Used synonymously with *measurement*.

Objective A statement of the desired result to be achieved within a specified time.

Occurrence An unusual or unplanned event having programmatic significance such that it adversely affects or potentially affects the performance, reliability, or safety of a facility.

Outliers Data that fall outside the control lines.

Parameter A quantity that describes a statistical population or any of a set of physical properties whose values determine the characteristics or behaviors of something.

Pareto analysis Also known as *distributing diagram*. A type of analysis that focuses attention on areas that have the most influence on the total, facilitating the assignment of resources in order to prioritize improvement efforts.

Performance based Being associated with the outcome rather than the process.

Performance goal The target level of outcomes expressed as a tangible, measurable objective against which actual achievements can be compared.

Performance indicator(s) A parameter useful for determining the degree to which an organization has achieved its goals.

Also, a quantifiable expression used to observe and track the status of a process.

Also, the operational information that is indicative of the performance or condition of a facility, group of facilities, or site.

Performance measure(s) Encompassing the quantitative basis by which objectives are established and performance is assessed and gauged. Includes *performance objectives and criteria* (POCs), performance indicators, and any other means that evaluate the success in achieving a specified goal.

Also, the quantitative results used to gauge the degree to which an organization has achieved its goals.

Performance objectives and criteria (POCs) The quantifiable goals and the basis by which the degree of success in achieving these goals is established.

Periodicity Data that show the same pattern of change over time, frequently seen in cyclical data.

Quality The degree to which a product or service meets customer requirements and expectations.

Quality assurance Actions that provide confidence that quality is achieved.

Quality management The management of a process to maximize customer satisfaction at the lowest cost.

Root cause The basic reasons for conditions adverse to quality that, if corrected, will prevent occurrence or recurrence.

Root cause analysis An analysis performed to determine the cause of part, system, and component failures.

Runs Series of data points above or below the central line. A *run* of 7 consecutive points or 10 out of 11 points indicates an abnormality.

Self-assessment A systematic evaluation of an organization's performance, with the objective of finding opportunities for improvement and exceptional practices; normally performed by the people involved in the activity, but may also be performed by others within the organization with an arms-length relationship to the work processes.

Situation analysis The assessment of trends, strengths, weaknesses, opportunities, and threats, giving a picture of the organization's internal and external environment to determine the opportunities or obstacles to achieving organizational goals; performed in preparation for *strategic planning* efforts.

Special causes of variation Also known as *assignable causes of variation*. A cause that is specific to a group of workers, a particular worker, a specific machine, or a specific local condition. Examples are water in a gasoline tank or poor spark plugs.

Stakeholder Any group or individual who is affected by or who can affect the future of an organization, customers, employees, suppliers, owners, other agencies, Congress, and critics.

Standard deviation A statistic used as a measure of the dispersion in a distribution. The square root of the arithmetic average of the squares of the deviations from the mean.

Strategic planning A process for helping an organization envision what it hopes to accomplish in the future, identify and understand obstacles and opportunities that affect the organization's ability to achieve that vision, and set forth the plan of activities and resource use that will best enable the achievement of the goals and objectives.

Surveillance The act of monitoring or observing a process or activity to verify conformance to specified requirements.

Task A well-defined unit of work with an identifiable beginning and end that is a measurable component of the duties and responsibilities of a specific job.

Top down To start with the highest level of management in an organization and propagate through successively lower levels of the organization.

Trend analysis A statistical methodology used to detect net changes or trends in levels over time. An analysis of parts, systems, component surveillances, performance, and operating histories to determine such things as failure causes, operational effectiveness, cost-effectiveness, and other attributes.

Also, the continual rise or fall of data points. If seven data points rise or fall continuously, an abnormality is considered to exist.

U-chart Also referred to as a *rate chart*. These deal with event counts when the area of opportunity is not constant during each period. The steps to follow for constructing a U-chart are the same as a C-chart, except that the control limits are computed for each individual time period since the number of standards varies.

Unit of measure A defined amount of some quality feature that permits the evaluation of that feature in numbers.

Validation A determination that an improvement action is functioning as designed and has eliminated the specific issue for which it was designed. Also, to determine or test the truth or accuracy by comparison or reference.

Verification The determination that an improvement action has been implemented as designed.

Also, the process of evaluating hardware, software, data, or information to ensure compliance with stated requirements. The act of reviewing, inspecting, testing, checking, auditing, or otherwise determining and documenting whether items, processes, services, or documents conform to specified requirements.

Vertical axis scaling The following general criteria should be applied to the depiction of trend data on control charts:

1. The scale should be set so that the chart can be quickly understood.
2. Data together with the limit lines should span at least half of the vertical axis.

Work A process of performing a defined task or activity (e.g., research and development, operations, maintenance and repair, administration, software development and use).

X-chart Involves the analysis of individual measured quantities for indications of process control or unusual variation. The standard deviation for an X-chart (also referred to as an *individual chart*) is calculated using a moving range.

Appendix B: Basic work-related formulas and conversion factors

Activity-planning formulas

ES: earliest starting time
EC: earliest completion time
LS: latest starting time
LC: latest completion time
t: activity duration
T: project duration
n: number of activities in the project network

Earliest start time for activity i

$$\mathrm{ES}(i) = \underset{j \in P\{i\}}{\mathrm{Max}} \{\mathrm{EC}(j)\}$$

where $P\{i\}$ = the set of immediate predecessors of activity i.

Earliest completion time of activity i

$$\mathrm{EC}(i) = \mathrm{ES}(i) + t_i$$

Earliest completion time of a project

$$\mathrm{EC}(\mathrm{Project}) = \mathrm{EC}(n)$$

where n is the last node.

Latest completion time of a project

$$\text{LC}(\text{Project}) = \text{EC}(\text{Project})$$

if no external deadline is specified.

$$\text{LC}(\text{Project}) = \text{T}_p$$

if a desired deadline, T_p, is specified.

Latest completion time for activity i

$$\text{LC(i)} = \underset{j \in S\{i\}}{\text{Min}}\{\text{LS(j)}\}$$

where $S\{i\}$ = the set of immediate successors to activity i.

Latest start time for activity i

$$\text{LS(i)} = \text{LC}(i) - t_i$$

Total slack (TS)

$$\text{TS}(\text{i}) = \text{LC}(\text{i}) - \text{EC}(\text{i})$$

$$\text{TS}(\text{i}) = \text{LS}(\text{i}) - \text{ES}(\text{i})$$

Free slack (FS)

$$\text{FS(i)} = \underset{j \in S(\text{i})}{\text{Min}}\left\{\text{ES}(j)\right\} - \text{EC(i)}\}$$

Interfering slack (IS)

$$\text{IS}(\text{i}) = \text{TS}(\text{i}) - \text{FS}(\text{i})$$

Independent float (IF)

$$\text{IF}(i) = \text{Max}\left\{0, \left(\text{Min ES}_j - \text{Max LC}_k - t_i\right)\right\}$$

$$j \in S\{i\} \text{ and } k \in P\{i\}$$

Where ES_j = the earliest starting time of a succeeding activity from the set of successors of activity i, and LC_k = the latest completion time of a preceding activity from the set of predecessors of activity i.

Program evaluation and review technique (PERT) formulas

$$t_e = \frac{a + 4m + b}{6}$$

$$s^2 = \frac{(b - a)^2}{36}$$

where:

- a = an optimistic time estimate
- b = a pessimistic time estimate
- m = the most likely time estimate
- t_e = the expected time
- S^2 = the variance of expected time

Critical path method (CPM) computation

$$\lambda = \frac{\alpha_2 - \beta}{\alpha_2 - \alpha_1}(100\%)$$

where:

α_1 = the minimum total slack in the CPM network
α_2 = the maximum total slack in the CPM network
β = the total slack for the path whose criticality is to be calculated

Task weight

The work content of a project is expressed in terms of days.

$$\text{Task weight} = \frac{\text{work-days for activity}}{\text{total work-days for project}}$$

Expected percent completion

$$\text{Expected \% completion} = \frac{\text{work-days for acti}}{\text{work-days plann}}$$

Expected relative percent completion

$$\text{Expected relative \% completion} = (\text{expected \% completion}).(\text{task weight})$$

Actual relative percent completion

$$\text{Actual relative \% completion} = (\text{actual \% completion}).(\text{task weight})$$

Planned project percent completion

$$\text{Planned project \% completion} = \frac{\text{work-days completed on project}}{\text{total work-days planned}}$$

Project-tracking index

$$\text{Project-tracking index} = \frac{\text{actual relative \% completion}}{\text{expected relative \% completion}} - 1$$

Permutations

$$P(n,m)=\frac{n!}{(n-m)!},\quad (n\geq m)$$

Combinations

$$C(n,m)=\frac{n!}{m!(n-m)!},\quad (n\geq m)$$

Failure rate

$$q=1-p=\frac{n-s}{n}$$

Mechanical advantage (MA)

The ratio of the force of resistance to the force of effort.

$$MA=\frac{F_R}{F_E}$$

Mechanical advantage formula for levers

$$F_R\cdot L_R=F_E\cdot L_E$$

Mechanical advantage formula for axles

$$MA_{\text{wheel and axle}}=\frac{r_E}{r_R}$$

$$F_R\cdot r_R\cdot=F_E\cdot r_E$$

where r_R = the radius (m) of the resistance wheel and r_E = the radius (m) of effort wheel.

Mechanical advantage formula for pulleys

$$MA_{\text{pulley}} = \frac{F_R}{F_E} = \frac{nT}{T} = n$$

where T = the tension in each supporting strand and n = the number of strands holding the resistance.

Mechanical advantage formula for inclined planes

$$MA_{\text{inclined plane}} = \frac{F_R}{F_E} = \frac{l}{h}$$

where l = the length (m) of the plane and h = the height (m) of the plane.

Mechanical advantage formula for wedges

$$MA = \frac{s}{T}$$

where s = the length of either slope and T = the thickness of the longer end.

Mechanical advantage formula for screws

$$MA_{\text{screw}} = \frac{F_R}{F_E} = \frac{U_E}{h}$$

$$F_R \cdot h = F_E \cdot U_E$$

where h = the pitch of the screw and U_E = the circumference of the handle of the screw.

Distance (S), speed (V), and time (t) formula

$$S = Vt$$

Newton's First Law of Motion

An object that is in motion continues in motion with the same velocity at constant speed and in a straight line, and an object at rest continues at rest unless an unbalanced (outside) force acts on it.

Newton's Second Law of Motion

The total force acting on an object equals the mass of the object times its acceleration.

$$F = ma$$

where:

F = total force
m = mass
a = acceleration

Newton's Third Law of Motion

For every force applied by object A to object B (action), there is a force exerted by object B on object A (the reaction) that has the same magnitude but is opposite in direction.

$$F_B = -F_A$$

where F_B = the force of action and F_A = the force of reaction.

Momentum of force

Momentum can be defined as mass in motion. Momentum is a vector quantity, for which direction is important.

$$p = mv$$

Conservation of momentum

In the absence of external forces, the total momentum of the system is constant. If two objects of mass m_1 and mass m_2, having velocity v_1 and v_2, collide and then separate with velocity v_1' and v_2', the equation for the conservation of momentum is

$$m_1 v_1 + m_2 v_2 = m_1 v_1 + m_2 v_2$$

Friction

$$F_f = \mu F_n$$

where:

F_f = the frictional force (N)
F_n = the normal force (N)
μ = the coefficient of friction ($\mu = \tan \alpha$)

Gravity

Gravity is a force that attracts bodies of matter toward each other. Gravity is the attraction between any two objects that have mass.

$$F = \Gamma \frac{m_A m_B}{r^2}$$

where:

m_A, m_B = the mass of objects A and B
F = the magnitude of attractive force between objects A and B
r = the distance between objects A and B
Γ = the gravitational constant ($\Gamma = 6.67 \times 10^{-11}$ N m²/kg²)

Gravitational force

$$F_G = g \frac{R_e^2 m}{(R_e + h)^2}$$

On the Earth's surface, $h = 0$; therefore,

$$F_G = mg$$

where:

F_G = the force of gravity
R_e = the radius of the Earth ($R_e = 6.37 \times 10^6$ m)
m = mass
g = acceleration due to gravity
g = 9.81 (m/s²) or 32.2 (ft/s²)

The acceleration of a falling body is independent of the mass of the object. The weight F_w on an object is actually the force of gravity on that object.

$$F_w = mg$$

Centrifugal force

$$F_c = \frac{mv^2}{r} = m\omega^2 r$$

Centripetal force

$$F_{cp} = -F_c = \frac{mv^2}{r}$$

where:

- Fc = centrifugal force (N)
- F_{cp} = centripetal force (N)
- m = the mass of the body (kg)
- v = the velocity of the body (m/s)
- r = the radius of the curvature of the path of the body (m)
- ω = angular velocity (s^{-1})

Torque

$$T = F \cdot I$$

where:

- T = torque (N m or lb ft)
- F = applied force (N or lb)
- I = the length of the torque arm (m or ft)

Work

Work is the product of a force in the direction of the motion and the displacement.

a. *Work done by a constant force*

$$W = F_s \cdot s = F \cdot s \cdot \cos\alpha$$

where:

W = work (Nm = J)
F_s = the component of force along the direction of movement (N)
s = the distance the system is displaced (m)

b. *Work done by a variable force*

If the force is not constant along the path of the object, then

$$W = \int_{si}^{sf} F_s(s) \cdot ds = \int_{si}^{sf} F(s)\cos\alpha \cdot ds$$

where:

$F_s(s)$ = the component of the force function along the direction of movement
$F(s)$ = the function of the magnitude of the force vector along the displacement curve
S_i = the initial location of the body
S_f = the final location of the body
α = the angle between the displacement and the force

Energy

Energy is defined as the ability to do work.

$$TME_i + W_{ext} = TME_f$$

where:

TME_i = the initial amount of total mechanical energy (J)
W_{ext} = the work done by external forces (J)
TME_f = the final amount of total mechanical energy (J)

a. *Kinetic energy*

Kinetic energy is the energy due to motion.

$$E_k = \frac{1}{2}mv^2$$

where m = mass of moving object (kg) and v = velocity of moving object (m/s).

b. *Potential energy*

Potential energy is the stored energy of a body and is due to its position.

$$E_{pg} = m \cdot g \cdot h$$

where:

E_{pg} = the gravitational potential energy (J)
m = the mass of the object (kg)
h = the height above reference level (m)
g = the acceleration due to gravity (m/s^2)

Conservation of energy

In any isolated system, energy can be transformed from one kind to another, but the total amount of energy is constant (conserved).

$$E = E_k + E_p + E_e + \ldots = \text{constant}$$

The conservation of mechanical energy

$$E_k + E_p = \text{constant}$$

Power

Power is the rate at which work is done or the rate at which energy is transformed from one form to another.

$$P = \frac{W}{t}$$

where:

P = power (W)
W = work (J)
t = time (s)

Since the expression for work is $W = F \cdot s$, the expression for power can be rewritten as

$$P = F \cdot v$$

where s = displacement (m) and v = speed (m/s).

Numbers and prefixes

Yotta (10^{24})	1 000 000 000 000 000 000 000 000
Zetta (10^{21})	1 000 000 000 000 000 000 000
Exa (10^{18})	1 000 000 000 000 000 000
Peta (10^{15})	1 000 000 000 000 000
Tera (10^{12})	1 000 000 000 000
Giga (10^{9})	1 000 000 000
Mega (10^{6})	1 000 000
Kilo (10^{3}):	1 000
Hecto (10^{2})	100
Deca (10^{1})	10
Deci (10^{-1})	0.1
Centi (10^{-2})	0.01
Milli (10^{-3})	0.001
Micro (10^{-6})	0.000 001
Nano (10^{-9})	0.000 000 001
Pico (10^{-12})	0.000 000 000 001
Femto (10^{-15})	0.000 000 000 000 001
Atto (10^{-18})	0.000 000 000 000 000 001
Zepto (10^{-21})	0.000 000 000 000 000 000 001
Yacto (10^{-24})	0.000 000 000 000 000 000 000 001
Stringo (10^{-35})	0.000 000 000 000 000 000 000 000 000 000 000 01

Constants

Speed of light	2.997925×10^{10} cm/s
	983.6×10^{6} ft/s
	186,284 mi/s
Velocity of sound	340.3 m/s
	1,116 ft/s
Gravity	9.80665 m/s^2
(Acceleration)	32.174 ft/s^2
	386.089 in/s^2

Area

Multiply	By	To obtain
Acres	43,560	Sq. feet
	4,047	Sq. meters
	4,840	Sq. yards
	0.405	Hectare
Sq. centimeters	0.155	Sq. inches
Sq. feet	144	Sq. inches
	0.09290	Sq. meters
	0.1111	Sq. yards
Sq. inches	645.16	Sq. millimeters
Sq. kilometers	0.3861	Sq. miles
Sq. meters	10.764	Sq. feet
	1.196	Sq. yards
Sq. miles	640	Acres
	2.590	Sq. kilometers

Volume

Multiply	By	To obtain
Acre-feet	1233.5	Cubic meters
Cubic centimeters	0.06102	Cubic inches
Cubic feet	1728	Cubic inches
	7.480	Gallons (US)
	0.02832	Cubic meters
	0.03704	Cubic yards
Liter	1.057	Liquid quarts
	0.908	Dry quarts
	61.024	Cubic inches
Gallons (US)	231	Cubic inches
	3.7854	Liters
	4	Quarts
	0.833	British gallons
	128	US fluid ounces
Quarts (US)	0.9463	Liters

Energy and heat power

Multiply	By	To obtain
BTU	1055.9	Joules
	0.2520	Kilogram-calories
Watt-hours	3600	Joules
	3.409	BTU
HP (electric)	746	Watts
BTU/second	1055.9	Watts
Watt-seconds	1.00	Joules

Mass

Multiply	By	To obtain
Carat	0.200	Cubic grams
Grams	0.03527	Ounces
Kilograms	2.2046	Pounds
Ounces	28.350	Grams
Pounds	16	Ounces
	453.6	Grams
Stone (UK)	6.35	Kilograms
	14	Pounds
Tons (net)	907.2	Kilograms
	2000	Pounds
	0.893	Gross tons
	0.907	Metric tons
Tons (gross)	2240	Pounds
	1.12	Net tons
	1.016	Metric tons
Tonnes (metric)	2,204.623	Pounds
	0.984	Gross pounds
	1000	Kilograms

Temperature

Conversion formulas	
Celsius to Kelvin	$K = C + 273.15$
Celsius to Fahrenheit	$F = (9/5)C + 32$
Fahrenheit to Celsius	$C = (5/9)(F - 32)$
Fahrenheit to Kelvin	$K = (5/9)(F + 459.67)$
Fahrenheit to Rankin	$R = F + 459.67$
Rankin to Kelvin	$K = (5/9)R$

Velocity

Multiply	By	To obtain
Feet/minute	5.080	Millimeters/second
Feet/second	0.3048	Meters/second
Inches/second	0.0254	Meters/second
Kilometers/hour	0.6214	Miles/hour
Meters/second	3.2808	Feet/second
	2.237	Miles/hour
Miles/hour	88.0	Feet/minute
	0.44704	Meters/second
	1.6093	Kilometers/hour
	0.8684	Knots
Knots	1.151	Miles/hour

Pressure

Multiply	By	To obtain
Atmospheres	1.01325	Bars
	33.90	Feet of water
	29.92	Inches of mercury
	760.0	Millimeters of mercury
Bar	75.01	Centimeters of mercury
	14.50	Pounds/sq. inch
Dyne/sq. centimeter	0.1	Newtons/sq. meter
Newtons/sq. centimeter	1.450	Pounds/sq. inch
Pounds/sq. inch	0.06805	Atmospheres
	2.036	Inches of mercury
	27.708	Inches of water
	68.948	Millibars
	51.72	Millimeters of mercury

Distance

Multiply	By	To obtain
Angstrom	10^{-10}	Meters
Feet	0.30480	Meters
	12	Inches
Inches	25.40	Millimeters
	0.02540	Meters
	0.08333	Feet
Kilometers	3280.8	Feet
	0.6214	Miles
	1094	Yards
Meters	39.370	Inches
	3.2808	Feet
	1.094	Yards
Miles	5280	Feet
	1.6093	Kilometers
	0.8694	Nautical miles
Millimeters	0.03937	Inches
Nautical miles	6076	Feet
	1.852	Kilometers
Yards	0.9144	Meters
	3	Feet
	36	Inches

Index

D

E

F

G

Y